A FLORA OF THE TRINITY ALPS OF NORTHERN CALIFORNIA

A Flora of the Trinity Alps of Northern California

by WILLIAM J. FERLATTE

with illustrations by Charles S. Papp

UNIVERSITY OF CALIFORNIA PRESS

Berkeley • Los Angeles • London

University of California Press
Berkeley and Los Angeles, California

University of California Press, Ltd.
London, England

ISBN 0-520-02089-8
Library of Congress Catalog Card Number: 72-635566
Printed in the United States of America

For Nicholas and others
like him who will learn to find
peace and beauty in nature.

Contents

Acknowledgments

I would like to thank Dr. Dennis Anderson for first suggesting this project and for providing assistance on many occasions. Dr. Daniel Norris and Dr. John Young also made many useful suggestions. The following people provided assistance with certain plant groups: Dr. Rimo Bacigalupi, *Mimulus*; Dr. Lincoln Constance, Umbelliferae; Lauramay Dempster, *Galium*; J. T. Howell and Hans Leschke, *Carex*; Dr. Richard Spellenberg, *Panicum*; and Dr. John Strother, Compositae. The herbaria of the University of California, Berkeley and Humboldt State University provided assistance with the use of their facilities. I am grateful to Alice Howard for doing a difficult yet excellent job in typing the manuscript. Thanks are due to Carl Bontrager and Blaine Rogers for their assistance in the field. John Strother made many critical and useful comments and his photographic contributions are appreciated. I would like to thank Charles S. Papp for his advice and fine illustrations. There are many other people who helped on numerous occasions and I wish to express my gratitude to them. Finally, I would like to thank Dr. Louise Watson and Mr. and Mrs. Gil Gates of Mountain Meadow Ranch. Without the help and continuing interest of these people my task would have been much more difficult and much less enjoyable.

Introduction

The Trinity Alps have intrigued botanists for many years, their rugged peaks and canyons remaining elusive except to a few hardy collectors. The presence of a predominantly Sierran flora only sixty miles from the Pacific Ocean is unique in California. In order to appreciate fully the floristic diversity found here one must forsake automobile and horse to travel on foot, away from the beaten path. Pristine niches may be found as they must have existed centuries ago, slow to change, essentially unspoiled by the activities of man.

It is hoped that this small book will increase the enjoyment of those who come to the Trinities seeking their natural beauty. It is also hoped that professional ecologists, foresters, and other biologists will find this work useful as a basic tool for further research into the natural history of the Trinity Alps.

PHYSIOGRAPHIC DESCRIPTION OF THE STUDY AREA

The Trinity Alps form a magnificent block of scenery about fifty miles west of Redding, California (lat. 41° N, long. 123° W), in the southern part of the Klamath region. Elevations in the Trinities range from 2,400 feet in the canyon of Stuart Fork to 9,002 feet at the summit of Thompson Peak (map). The region is drained by the South Fork of the Salmon River and the Trinity River. The Trinity has several important tributaries which head in the interior regions of the Alps: the North Fork, Stuart Fork, Canyon Creek, and Coffee Creek. Except for a few private holdings, most of the Alps are included in the Salmon-Trinity Primitive Area.

Precipitation in northwestern California ranges from thirty to fifty inches with some local areas receiving as much as one hundred inches annually (Irwin, 1960). Summer rainfall is infrequent (MacGinitie, 1937) but occasional thundershowers can be expected in the high mountains. Snow may fall on the higher peaks any time between October and June. The only meteorological records made in the Trinity Alps proper are snow pack surveys compiled by the U.S. Forest Service for the State Department of Water Resources. Readings taken in April of each year from 1946 through 1967 indicate an average

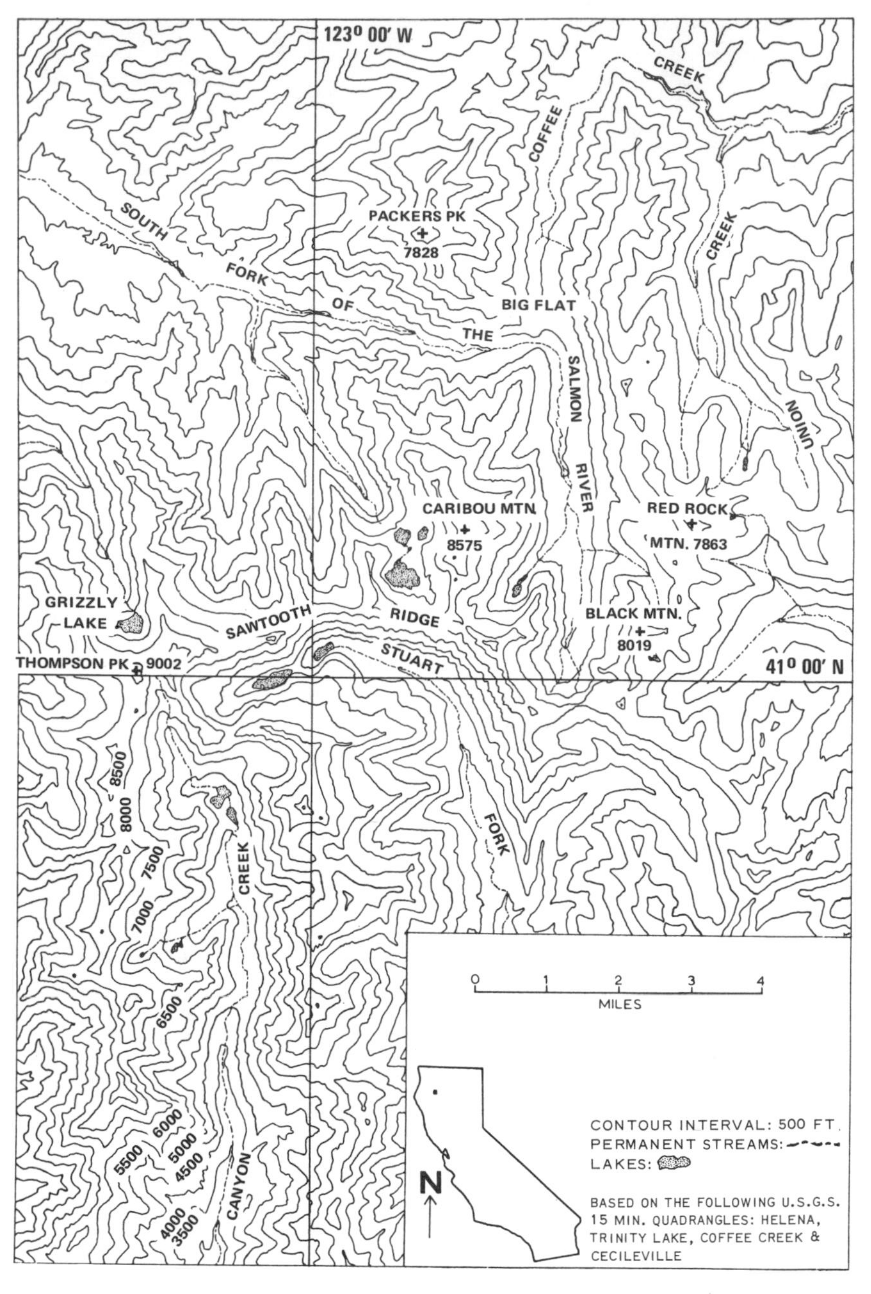

THE TRINITY ALPS

snow depth of 31.4 inches with an average water content of 13.0 inches for Big Flat at an elevation of 5,100 feet and an average depth of 85.5 inches with an average water content of 38.4 inches for the south side of Red Rock Mountain at an elevation of 6,600 feet (State Department of Water Resources, Basic Data Supplement for April, Bulletin 120, 1962–67, and Summary of California Snow Survey Measurements Through 1959). The growing season lasts from May to September at Big Flat and is proportionately shorter at higher elevations, probably limited to four or five weeks on the north side of Thompson Peak above 8,000 feet.

In the latter part of the nineteenth century and the early twentieth the Trinity Alps were vigorously mined for gold and other metallic deposits. Hard rock shaft and tunnel mining as well as placer mining was practiced. It was the latter that made the greatest visible effect on the present landscape as illustrated by the dredge tailings on upper Coffee Creek.

The major rock units in the Trinities have been mapped by Davis, et al. (1965). The oldest rocks, a series of Pre-Upper Jurassic metamorphics, are divided into the following formations based on structural relationships and mineral composition: the Grouse Ridge Formation consisting mainly of micaceous quartzite, amphibolites, calc-schists, and impure marbles; the Salmon Formation consisting of albite-epidote-hornblende schists; and the Stuart Fork Formation consisting of metachert, micaceous quartzite, minor marbles, and mica, greenstone, and graphite phyllites. Igneous rocks are of two major types and are mostly Upper Jurassic-Lower Cretaceous in age. Canyon Creek and Caribou Mountain plutons are composed of quartz diorite and there are also large outcrops of ultramafic rocks most of which have been altered to serpentine (Irwin, 1960). The youngest rocks are Quaternary glacial deposits, alluvium, and talus.

Several important geological events were responsible for the present mountains with their associated drainage patterns and mineral deposits. During a short interval of late Jurassic time, rocks in the Trinity Alps region were folded, metamorphosed, and intruded by ultramafic and granitic plutons (Lipman, 1962). The intrusive process apparently was relatively slow and the plutons were subjected to the regional tectonic forces that were acting during the time they were emplaced, as is suggested by certain structural trends in the plutons being concordant with similar trends in neighboring country rock. This period of tectonism may be related, at least in time, to the formation of the ancestral Sierra Nevada.

The present high regions of the Alps are regarded as projections above remnants of the Klamath peneplain (Irwin, 1960), which was formed in the late Cretaceous-early Tertiary (MacGinitie, 1937) by a period of erosion that resulted in a landscape of relatively low relief with a maximum elevation of perhaps 2,500 feet. There was probably little tectonic activity in the area at this time and it may be that the late Jurassic intrusives were initially exposed by gradual erosion during formation of the peneplain.

In the late Pliocene-early Pleistocene (MacGinitie, 1937) a major uplift began on a regional scale and a new, invigorated cycle of erosion ensued. Because of this uplift—a westward tilting resulting in higher regions to the east

(Irwin, 1960)—antecedent streams were rejuvenated and began to erode the steep, narrow, youthful canyons now seen in the area. This westward tilting may account for the fact that the overall drainage pattern in the Klamath region is more or less transverse to the prevailing north-south trending grain of the mountains. However, there are some instances in the Trinity Alps where streams in fact do follow the structural grain of the region. The North Fork of the Trinity River trends in an almost due north-south direction. The upper reaches of the South Fork of the Salmon River, Coffee Creek, and Trinity River also have similar trends. The courses of Canyon Creek and Stuart Fork are possibly controlled by structural folds in the Canyon Creek pluton and adjacent metamorphics.

During formation of the Canyon Creek pluton in the late Jurassic, concurrent regional folding exerted its effect on the pluton as it was forcibly intruded, producing a series of northwest trending folds in the granitic mass (Lipman, 1962). Canyon Creek, which heads in the central area of the Canyon Creek pluton, occupies a broad structural trough that developed at the time of intrusion. The creek flows in a southerly direction and more or less conforms with the local structural trend. In the next drainage system to the east, Stuart Fork follows the trend of the Stuart Fork Antiform, a structural feature of the Stuart Fork Formation, for some distance. According to Lipman's (1962) interpretation the forceful intrusion of the Canyon Creek pluton caused the neighboring rocks to bow out around it. Stuart Fork also flows around the pluton. In its upper course where it flows out of Emerald Lake the stream follows the contact of the pluton to a remarkable degree. It is not until the stream reaches lower elevations and flows into Trinity Lake that it breaks away from the structural trend. Although these exceptions hardly take away from the fact that in general the regional drainage pattern is not under structural control, it seems that the structural grain has had a more important local effect in the Trinity Alps.

The latest period of uplift must have been a factor in causing the higher elevations of the Alps to be subjected to glaciation in the Pleistocene. Compared to the Sierra Nevada, glacial activity was relatively limited in the Alps; nevertheless, ample evidence of glaciation can be seen in the topography, the location of streams and lakes, and certain soil relationships.

Sharp (1960) has studied the glacial features of the Trinity Alps and finds evidence of at least four substages of glacial activity during the Pleistocene. Three of these substages belong to the Wisconsin stage and the earliest is regarded as pre-Wisconsin, perhaps as old as Kansan (Irwin, 1960). There may have been even earlier stages of glaciation as suggested from stream terraces in the lower reaches of Canyon Creek, but evidence is relatively meager. Sharp (1960) feels that there has been a recurrence of glacial activity at higher elevations since the end of Wisconsin time and calls this the Neoglacial stage. There are several very young, little disturbed terminal moraines on the northern sides of some of the higher peaks. Judging from the degree of weathering and the amount of vegetational cover they may range in age from a few hundred to 2,000 years. The last remaining permanent ice fields, located on the north side of Thompson Peak at elevations of 8,200 and 8,500 feet, cover five to six acres each. The fact that they are situated in the shadow of the highest ridge in the Trinity Alps and benefit from the relatively high north coastal precipitation probably accounts for their persistence.

Sharp (1960) believes there were sheets of ice two to three hundred feet thick that extended from the glaciers to the "schrund" lines below the ridge crests in the broad upper reaches of many canyons. He calls these sheets ice carapaces. The cliffs formed by "bergschrunding" are fifty to three hundred feet high at present and make travel at higher elevations difficult. From the base of these cliffs the scoured granite walls slope less steeply down to the canyon bottoms.

The mineralogy and resultant color variations of the igneous and metamorphic rocks in the Trinity Alps are often of a contrasting nature. Canyon Creek pluton, consisting primarily of a light colored quartz diorite, has intruded the Salmon Formation, a dark hornblende schist. Such differences in color and lithology make the tracing of glacial deposits back to their origin a relatively simple task (Sharp, 1960). Other obvious glacial features such as cirques, bedrock basins, marshy meadows, U-shaped valleys, erratics, and striations can be seen at higher elevations (Irwin, 1960).

There is a striking example of stream piracy in the Alps where the upper South Fork of the Salmon River has beheaded Coffee Creek at Big Flat. Sharp (1960) surmises that this occurred in early or middle Wisconsin and that it was a direct result of glaciation. As the glacier moved northward down the valley of Coffee Creek, marginal meltwaters spilled westward over the low divide between the Coffee Creek drainage and the headwaters of the South Fork of the Salmon River. This new stream cut rapidly into the low ridge as it fed into the steeper gorge and was able to maintain its course after the glacier receded. At the present time the upper South Fork of the Salmon River makes an abrupt westward turn and flows at a right angle to the trend of the valley only about one mile from where Coffee Creek now heads at the same elevation.

ECOLOGICAL ZONES

The higher elevations of the Trinity Alps can be divided into five major zones based on assemblages of characteristic species and different climatological and topographic factors. Minor habitats such as stream and lake banks, bogs, meadows, and talus slopes have been included within the larger zones. Each of these zones will be considered separately.

MIXED CONIFER FOREST

The Mixed Conifer Forest is found on the valley bottoms and lower slopes of adjacent ridges with elevations ranging from 5,000 to 6,000 feet. It is an area of overlap between high and low elevation species and contains the most varied flora of any of the ecological zones in the Alps. Snow has usually melted by early June and different species can be found in flower from May through September. *Viola sheltonii*, *Claytonia lanceolata*, and *Lewisia nevadensis* are common elements of the early flora, while *Orthocarpus copelandii* and *Haplopappus bloomeri* bloom in late summer. Other common species included in this zone are *Pinus ponderosa*, *P. lambertiana*, *P. contorta* ssp. *latifolia*, *Abies concolor*, *Pseudotsuga menziesii*, *Calocedrus decurrens*, *Adenocaulon bicolor*, *Amelanchier pallida*, *Spiraea douglasii*, *Clintonia uniflora*, and *Pyrola picta*.

RED FIR FOREST

The Red Fir Forest occupies essentially a transition zone with elevations ranging from 5,500 to 7,000 feet. Snow often remains until the end of June. The most obvious species here are *Abies magnifica* var. *shastensis*, *Pinus jeffreyi*, *P. monticola*, *Ribes nevadense*, *Leucothoe davisiae*, *Lupinus croceus*, *Veratrum viride*, and *Ligusticum californicum*.

SUBALPINE FOREST

The most common species of the Subalpine Forest are *Pinus balfouriana*, *P. albicaulis*, *Tsuga mertensiana*, *Arctostaphylos nevadensis*, *Anemone occidentalis*, *Kalmia polifolia* var. *microphylla*, *Phyllodoce empetriformis*, and *Cassiope mertensiana*. Elevations range from 7,000 to 8,500 feet and snow can last throughout the year in protected places. The subalpine forest occurs mainly along the summits of the higher ridges and in the higher basins of the Trinity Alps.

ALPINE FELL-FIELD

This zone is the smallest in area, usually developed on north-facing slopes above 8,000 feet. The growing season is short, reaching its peak in mid August. There are permanent ice fields in some places and there is ample melt water throughout the season. The following species are characteristic of the Alpine Fell-Fields: *Saxifraga tolmiei*, *Sibbaldia procumbens*, *Ranunculus eschscholtzii*, *Primula suffrutescens*, and *Oxyria digyna*.

MONTANE CHAPARRAL

This zone is extensively developed on the slopes above Coffee Creek, the South Fork of the Salmon River, and Canyon Creek at elevations from 5,000 to 7,000 feet. These slopes are well exposed and snow has often melted by early May. Soil is usually gravelly and well drained and there are few permanent streams. The most common shrubs in the chaparral are *Quercus vaccinifolia*, *Ceanothus velutinus*, *Arctostaphylos patula*, and *Garrya fremontii*.

DISCUSSION OF THE FLORA

The first record of a botanical collection from the Trinity Alps is that of Alice Eastwood (1902). H. M. Hall visited the Union Lake and Dorleska areas in 1909. Annie M. Alexander and Louise Kellogg made collections around Thompson Pk. in 1911 and Canyon Cr. and Stuart Fork in 1948. J. T. Howell collected in the Big Flat area and Caribou Basin in July, 1937. Other botanists have visited the Trinities from time to time but relatively little has been published on the flora up until now. Joseph P. Tracy apparently never collected in the Trinity Alps although he visited adjacent areas on numerous occasions.

Field work for this study was begun in November, 1965, and continued through the summer of 1973. Only elevations above 5,000 feet are considered; many species found at lower elevations are therefore excluded from this treatment. A summary of the divisions and families included in the flora, along with the numbers of genera and species in each family, is presented in Table I. Unless otherwise noted the species described were identified in *A California*

Table 1

A STATISTICAL SUMMARY OF THE FLORA

DIVISION	FAMILY	NUMBER OF GENERA	NUMBER OF SPECIES
Lepidophyta			
	Isoetaceae	1	2
	Selaginellaceae	1	1
Calamophyta			
	Equisetaceae	1	2
Pterophyta			
	Aspidiaceae	3	3
	Marsileaceae	1	1
	Ophioglossaceae	1	1
	Pteridaceae	5	5
Coniferophyta			
	Cupressaceae	2	2
	Pinaceae	5	13
	Taxaceae	1	1
Anthophyta			
	Aceraceae	1	2
	Amaryllidaceae	2	5
	Apocynaceae	1	1
	Aristolochiaceae	1	2
	Asclepiadaceae	1	2
	Berberidaceae	1	2
	Betulaceae	2	3
	Boraginaceae	4	8
	Campanulaceae	3	4
	Caprifoliaceae	3	6
	Caryophyllaceae	5	13
	Chenopodiaceae	1	1
	Celastraceae	1	1
	Compositae	27	57
	Convolvulaceae	1	2
	Cornaceae	1	2
	Crassulaceae	1	3
	Cruciferae	14	25
	Cuscutaceae	1	1
	Cyperaceae	2	29
	Ericaceae	8	9
	Euphorbiaceae	1	2
	Fagaceae	2	5

Table 1 (Continued)

DIVISION	FAMILY	NUMBER OF GENERA	NUMBER OF SPECIES
	Fumariaceae	1	2
	Garryaceae	1	1
	Gentianaceae	1	3
	Gramineae	19	43
	Hydrophyllaceae	4	9
	Hypericaceae	1	2
	Iridaceae	1	1
	Juncaceae	2	14
	Labiatae	6	7
	Leguminosae	4	15
	Liliaceae	14	17
	Linaceae	1	1
	Loasaceae	1	1
	Loranthaceae	1	1
	Malvaceae	2	2
	Nymphaeaceae	1	1
	Onagraceae	4	13
	Orchidaceae	7	11
	Orobanchaceae	2	2
	Plantaginaceae	1	1
	Polemoniaceae	8	14
	Polygalaceae	1	1
	Polygonaceae	4	17
	Portulacaceae	4	12
	Primulaceae	2	4
	Pyrolaceae	3	7
	Ranunculaceae	9	17
	Rhamnaceae	2	5
	Rosaceae	16	28
	Rubiaceae	2	6
	Salicaceae	2	5
	Sarraceniaceae	1	1
	Saxifragaceae	10	30
	Scrophulariaceae	9	38
	Solanaceae	1	1
	Sparganiaceae	1	1
	Umbelliferae	8	10
	Valerianaceae	2	2
	Verbenaceae	1	1
	Violaceae	1	7
Totals	73	266	571

flora (Munz, 1959), *An illustrated flora of the Pacific states,* (Abrams, 1940; 1944; 1951; Abrams and Ferris, 1960) and *Manual of the grasses of the United States* (Hitchcock, 1950). Identifications are based on written descriptions and herbarium specimens. The descriptions and keys are based primarily on the author's collections and observations in the field and herbarium. A complete set of specimens has been deposited in the Humboldt State University Herbarium (HSC), Arcata, California.

The following species were found to be endemic to the Klamath region: *Draba howellii, Erigeron cervinus, Heuchera merriamii, Lupinus croceus, Penstemon tracyi, Picea breweriana,* and *Raillardella pringlei.*

For the most part those species regarded as introduced and weedy are found around human habitations and in meadows that have been rather heavily grazed by cattle and horses. The introduced species include *Agrostis alba, Capsella bursa-pastoris, Cerastium vulgatum, Holcus lanatus, Hypericum perforatum, Lepidium campestre, Malva neglecta, Phleum pratense, Plantago major, Poa bulbosa, P. compressa, Rumex angiocarpus, Taraxacum officinale, Trifolium pratense,* and *Veronica serpyllifolia* var. *humifusa.*

The following species were previously not reported for the Trinity Alps: *Arnica amplexicaulis, A. diversifolia, A. mollis, A. viscosa, Artemisia norvegica* var. *saxatilis, Carex proposita, C. scopulorum, Claytonia nevadensis, Haplopappus lyallii, Juncus drummondii, Lithophragma parviflora, Lotus nevadensis, L. oblongifolius, Lupinus lyallii, Luzula divaricata, Melica fugax, Mimulus tilingii, Poa epilis, Raillardella pringlei, Sambucus melanocarpa, Saxifraga bryophora,* and *Sellaginella watsonii.*

The flora of the Trinity Alps is generally typical of the high mountains of California; 77% of the species collected occur in the Sierra Nevada; and 84% extend at least to Oregon with many species reaching as far north as Alaska. There is some influence from the east with 17% occurring in northeastern California and 13% ranging as far as the Rocky Mountains; 12% reach the southern limits of their range in northwestern California without occurring in the Sierra Nevada; and 8% of the species collected are restricted to California. The presence of species like *Arnica viscosa, Claytonia nevadensis, Haplopappus lyallii, Picea breweriana,* and *Pinus balfouriana* suggests that some elements of the flora may represent relicts of older populations that existed in past, cooler climates.

The phylogenetic sequence of the higher taxa and, unless otherwise noted, the nomenclature is that followed by Munz (1959, 1968). Common names are mainly those used in Abrams (1940; 1944; 1951; Abrams and Ferris, 1960). Place names are from the following U.S.G.S. topographic maps (15′ quadrangles): Helena, Cecilville, Coffee Creek, and Trinity Lake. Relative abundance, habitat, site of collection, the author's collection number or other collector's name and number and inclusive blooming dates follow each description. A list of additional species collected or observed is included after the last description.

Abbreviations and Symbols Used

ca.	circa (about or approximately)
cf.	compare
cm	centimeter, centimeters
Co.	county
Cr.	creek
diam.	diameter
dm	decimeter, decimeters
elev.	elevation
F.	forest
fl., fls.	flower, flowers
fld.	flowered
fr.	fruit
ft.	foot, feet
infl.	inflorescence
invol.	involucre
L.	lake
lf.	leaf
lft., lfts.	leaflet, leaflets
lvd.	leaved
lvs.	leaves
m	meter, meters
mm	millimeter, millimeters
Mt., Mts., mts.	mountain, mountains
Pk.	peak
R.	river
sp.	species
ssp.	subspecies
Tr.	trail
var.	variety
±	more or less

Keys and Descriptions

DIVISION LEPIDOPHYTA

Stems reduced to short, underground corms; lvs. tufted, slender, 4–20 cm long .... Isoetaceae p. 11
Stems well developed, prostrate, above ground; lvs. subulate, 2–3 mm long .... Selaginellaceae p. 12

ISOETACEAE

Isoetes L.

Megaspores covered with low, blunt tubercles .... *I. howellii*
Megaspores covered with rather long, sharp spines .... *I. muricata* var. *hesperia*

I. howellii Engelm. Howell's Quillwort. Perennial aquatic herb; corms covered with roots and lf. bases; lvs. 10–25 in number, 8–20 cm long, narrow, with membranous margins below, tapering to slender tips; ligule deltoid, 1.5–2 mm long; velum incomplete, covering one-third to one-half of the sporangium; megaspores 0.4–0.5 mm in diam., pale ashen gray when dry, tan when wet, covered with low, closely spaced tubercles which may be joined to form short discontinuous ridges, commissural ridges prominent; microspores 0.025–0.037 mm long, ± roughened. Found at one location. Edge of lake in water 3–12 dm deep, 5,800 ft. Red Fir F. Josephine L. (*281*). August.

I. muricata Durieu var. *hesperia* Reed. Braun's Quillwort. Perennial aquatic herb; corms covered with roots and lf. bases; lvs. 8–15 in number, 4–10 cm long, narrow, with broad, membranous margins below, tapering to slender tips; ligule narrow-triangular, 1–1.5 mm long; velum incomplete, covering one-quarter to one-third of the sporangium; megaspores 0.4–0.5 mm wide, white to pale tan, covered with fine spines, the commissural ridges somewhat obscured by the spines; microspores 0.020–0.030 mm long, tending to be smooth. Found at one location. Edge of pond, in water ca. 15 cm deep, 5,800 ft. Mixed Conifer F. Boulder Cr. L. (*470*). August.

SELAGINELLACEAE

Selaginella Beauv.

S. watsonii Underw. Alpine Selaginella. Densely cespitose perennial; stems prostrate, freely rooting, with numerous short, ascending branches; lvs. 2–3 mm long, subulate, with a prominent dorsal groove, cilia 4–5 or lacking, stout, the seta one-quarter to one-third the length of the lf., smooth, white-hyaline; strobili numerous, 1.5–3 cm long, strongly quadrangular; sporophylls ovate with a few stiff cilia toward the base; megaspores pale orange to yellow, 0.4–0.5 mm in diam., rugose-reticulate; microspores dark orange, 0.03–0.05 mm in diam., borne above and below the megaspores. Found at one location. Rock outcrop, 5,600 ft. Mixed Conifer F. Josephine L. Basin (*78*). June. This species has previously been reported in California only from the Sierra Nevada and White Mts. (Munz, 1959).

DIVISION CALAMOPHYTA

EQUISETACEAE

Equisetum L.

E. arvense L. Common Horsetail. Sterile stems green, spreading to erect, 10–20 cm long, the many slender branches borne in dense whorls; fertile stems flesh colored, 5–25 cm high, bearing strobili 2–3 cm long. Occasional. Open areas at edge of ponds and springs, 5,000 ft. Mixed Conifer F. Big Flat (*381*). August. Fertile stems were not observed during this study.

DIVISION PTEROPHYTA

Sporangia large, borne in a panicle; annulus absent.
Ophioglossaceae p. 14

Sporangia minute, borne on the backs of the lf. segments; annulus present.

Sori marginal, covered by the reflexed lf. margin. . Pteridaceae p. 13

Sori not marginal, covered by a true indusium or naked.
Aspidiaceae p. 12

ASPIDIACEAE

Sori naked . . . . *Athyrium*

Sori covered by indusia, at least when young.

Indusium attached along one side, hoodlike; delicate ferns.
Cystopteris

Indusium centrally attached, umbrellalike; coarse ferns.
Polystichum

Athyrium Roth

A. distentifolium Tausch ex Opiz var. *americanum* (Butters) Cronq. Alpine Lady-fern. Rhizome branched, forming rounded masses; lvs. 2–6 dm long, narrow, pinnately divided, the divisions distant; stipes short, narrow to somewhat thickened, sparsely chaffy with brown scales; sori many, ± rounded, lacking indusia. Common. Rocky slopes, 6,000–8,000 ft. Red Fir F. Grizzly L. Basin (*899*), Kidd Cr. (*1097*). The nomenclature here follows Hitchcock, et al. (1969).

Cystopteris Bernh.

C. fragilis (L.) Bernh. Brittle-fern. Delicate ferns from slender, matted rhizomes; lvs. 5–20 cm long including the slender stipe, glabrous, the blades narrow, pinnately divided, the ultimate segments serrate; sori scattered on the back of the segments and covered by a hoodlike indusium attached along one side which withers in age exposing the sporangia. Occasional. Shaded, rocky stream banks, 6,000–7,800 ft. Red Fir F. Kidd Cr. (*699*), Caribou Basin (*1381*).

Polystichum Roth

P. lemmonii Underw. Lemmon's Shield-fern. Coarse ferns from stout, tufted rhizomes; lvs. 1–5 dm long including the stipe, the blades pinnately divided, narrow; stipes thick, ± covered with brownish scales; sori scattered on the lower surface of the lf. segments, covered by round, centrally attached indusia. Common. Rocky slopes, 6,000–7,000 ft. Red Fir F. Head of Bullards Basin (*634*).

PTERIDACEAE

Lvs. dimorphous, fertile blades with linear divisions. . . . *Cryptogramma*
Lvs. not dimorphous, divisions of fertile blades as broad as those of sterile blades.
Reflexed margins of lf. segments continuous.
Lvs. 3–10 dm high; stipes straw colored. . . . . *Pteridium*
Lvs. mostly less than 2 dm high; stipes dark brown
Cheilanthes
Reflexed margins of lf. segments not continuous. . . . . *Adiantum*

Adiantum L.

A. pedatum L. var. *aleuticum* Rupr. Western Maiden-hair. Rhizomes stout, thick; lvs. up to 5 dm long including the blades, rounded in outline, 1–3 dm broad, forked at the base, the branches bearing several pinnately divided segments, the upper margins of the ultimate segments deeply cleft; stipe dark brown, glabrous, wiry; sori covered by the discontinuous, reflexed margins of the ultimate lf. segments. Occasional. Wet places, 5,000–7,000 ft. Mixed Conifer F., Red Fir F. Head of Bullards Basin (972).

Cheilanthes Sw.

C. gracillima D. C. Eat. in Torr. Lace-fern. Rhizome densely tufted and covered with narrow brown scales ca. 2 mm long; lf. blades 5–10 cm long, linear to lanceolate, twice pinnate, the ultimate segments oblong, 1–1.5 mm wide, the upper surface glabrous and yellow-green to green, the lower surface densely covered with pale to rusty brown scales and hairs, especially when young; stipes dark brown and wiry, soon becoming glabrous; sori borne at and covered by the inrolled margins of the lf. segments. Common. Cracks in rock walls and ledges, 5,000–8,000 ft. Mixed Conifer F., Red Fir F. Josephine L. Basin (*117, 601*).

Cryptogramma R. Br. in Richards.

C. acrostichoides R. Br. in Richards. American Parsley-fern. Rhizomes densely tufted, covered with narrow brown scales; lvs. twice pinnate, 10–25 cm long including the stipe, dimorphous, the fertile blades with linear ultimate segments, sterile blades with the ultimate segments ovate, lanceolate or obovate, coarsely serrate; stipes wiry, straw-colored to green, glabrous; sori submarginal, covered by the continuous reflexed margins of the lf. segments. Common. Cracks in rock walls and ledges, 5,500–8,000 ft. Mixed Conifer F., Red Fir F. Josephine L. Basin (*118, 600*), Grizzly L. Basin (*900*).

Pteridium Scop.

P. aquilinum (L.) Kuhn. var. *pubescens* Underw. Western Bracken. Coarse ferns with straw colored stipes; lvs. 3–10 dm high, thrice pinnate into lanceolate divisions, sparingly pubescent, the inrolled margins of the ultimate segments continuous forming a false indusium. Common. Dry woods and meadows, 5,000–6,000 ft. Mixed Conifer F. Canyon Cr. Meadow (*1434*).

OPHIOGLOSSACEAE

Botrychium Sw.

B. multifidum (Gmel.) Rupr. ssp. *silaifolium* (Presl.) Clausen. California Grape-fern. Fleshy ferns from small rootstocks; lvs. broad, 10–25 cm long, ternately divided, glabrous; sporangia large, ca. 1 mm in diam., borne in a panicle. Found at one location. Wet edge of meadow, 5,500 ft. Mixed Conifer F. Canyon Cr. (*795*).

DIVISION CONIFEROPHYTA

Lvs. scalelike .... Cupressaceae p. 14
Lvs. needlelike or linear.
  Fruit a dry cone of several to many woody scales, each scale bearing 2 seeds .... Pinaceae p. 15
  Fruit berrylike.
    Trees, 3–20 m tall...................... Taxaceae p. 19
    Low shrubs, less than 1 m high.... (*Juniperus*) Cupressaceae

CUPRESSACEAE

Lvs. scalelike; fr. a woody cone........................*Calocedrus*
Lvs. needlelike; fr. berrylike...........................*Juniperus*

Calocedrus Kurz.

C. decurrens (Torr.) Florin. Incense-Cedar. Forest tree 26–35 m tall; bark 1–2.5 cm thick, cinnamon-brown, fibrous; lvs. scalelike, 4-ranked, 3–10 mm long; cones oblong, 2–2.5 cm long, pendulous, the scales in 3 pairs. Common. Valley floors and lower slopes, 5,000–6,000 ft. Mixed Conifer F. This species was not collected but was observed throughout the study area at lower elevations. May.

Juniperus L.

J. communis L. var. *saxatilis* Pall. Dwarf Juniper. Low, spreading shrub, 3–10 dm high; branches densely covered with linear, sharply pointed, shiny green, needlelike lvs. 5–10 mm long; male cones 3–6 mm long; berry blue with a white bloom, globose, ca. 7 mm in diam. Occasional. Rocky, wooded slopes, 6,000–8,000 ft. Red Fir F., Subalpine F. This species was not collected but is found in Caribou Basin and upper Canyon Cr.

PINACEAE

Cones erect on the branches, the scales falling separately at maturity; lvs. solitary, blunt at apex . . . . *Abies*

Cones reflexed or pendulous, the scales persistent; lvs. solitary or fascicled, acute to sharp at apex.

  Lvs. needlelike, in fascicles of 2–5, mostly over 3 cm long. . . . *Pinus*

  Lvs. linear, solitary, mostly less than 3 cm long.

    Branchlets roughened by persistent lf. bases; bracts of cone scales not exserted, acute at the tip.

      Lvs. sessile on the woody, stalklike bases; cones stalked, 7–10 cm long . . . . *Picea*

      Lvs. narrowed to a short petiole; cones sessile, 2.5–7.5 cm long . . . . *Tsuga*

    Branchlets not roughened by persistent lf. bases; bracts of cone scales conspicuously exserted with a long taillike appendage produced from the notched tip . . . . *Pseudotsuga*

Abies Mill.

Cones 6–12 cm long; bracts half as long as scales; lvs. flat; inner bark tan. *A. concolor*

Cones 10–20 cm long; bracts three-quarters as long or longer than the scales; lvs. ± quadrangular; inner bark dark reddish-brown.

  Bracts included. . . . . *A. magnifica*

  Bracts well exserted. . . . . *A. magnifica* var. *shastensis*

A. concolor (Gord. & Glend.) Lindl. White Fir. Forest tree 15–70 m tall; bark on older trees gray on the surface, tan to yellowish-brown on fresh cuts, deeply furrowed; branchlets subglabrous; lvs. 3–6 cm long, flattened, with a median keel on one surface; cones oblong-cylindric, 7–12 cm long; bracts scarcely half as long as the scales, with a short, slender point at the rounded end. Common. Valley floors and immediately adjacent slopes, 5,000–6,000 ft. Mixed Conifer F. This species was not collected but was observed throughout the study area. May–June.

A. magnifica A. Murr. Red Fir. Forest tree 20–60 m tall; bark on older trees gray to red-brown on surface, dark reddish-brown on fresh cuts, deeply furrowed; branchlets puberulent the first season, later becoming glabrous; lvs. 2–3.5 cm long, ±4-sided, ribbed on upper and lower surfaces; cones oblong-cylindrical, 15–20 cm long; bracts ca. three-quarters as long as scales, abruptly contracted to a narrow tip. Occasional. Slopes, ridges, and narrow canyons, 6,000–8,000 ft. Red Fir F. Packers Pk. (*562*). June.

var. *shastensis* Lemmon. Bracts of cones well exserted and ± broadened at the apex. Common. Slopes, ridges, and narrow canyons, 6,000–8,500 ft. Red Fir F., Subalpine F. North side of Caribou Mtn. (*1082*). This is apparently the predominant variety of *A. magnifica* in the Trinity Alps, having been observed throughout the study area. *A. magnifica* occurs at scattered locations in relatively small numbers.

Picea Dietr.

P. breweriana Wats. Weeping Spruce. Tree to 30 m tall; bark reddish-brown, broken into long, thin, appressed scales; lower branches with slender, pendulous branchlets; lvs. 2–2.5 cm long, spreading from all sides of the branchlets; cones oblong, 7–10 cm long, stalked; bracts ca. one-quarter as long as the scales, oblong, acute. Occasional. Rocky slopes and canyons, 5,000–8,000 ft. Mixed Conifer F., Subalpine F. Canyon Cr. Lakes (*838*); also observed around Thompson Pk., Caribou Basin, and Grizzly L. Basin.

Pinus L.

Lvs. in fascicles of 5.
- Cones cylindric, mostly 15–45 cm long.
  - Cones 25–45 cm long; lvs. sharp-pointed...... *P. lambertiana*
  - Cones 10–25 cm long; lvs. obtuse............... *P. monticola*
- Cones ovoid to subglobose, 3.5–7.5 cm long.
  - Cone scales with terminal unarmed umbos ...... *P. albicaulis*
  - Cone scales with dorsal umbos armed with slender prickles. *P. balfouriana*

Lvs. in fascicles of 2 or 3.
- Lvs. in twos, 3–6 cm long .... *P. contorta* ssp. *latifolia*
- Lvs. in threes, 8–25 cm long.
  - Cones asymmetrical, remaining closed at maturity, placed laterally on the branches .... *P. attenuata*

Cones symmetrical, opening at maturity, placed subterminally on the branches.

Cones 7–15 cm long, narrowly ovoid, the scales armed with outwardly protruding prickles; bark lacking vanilla odor . . . . *P. ponderosa*

Cones 15–25 cm long, broadly ovoid, the scales armed with deflexed prickles; bark often having the odor of vanilla . . . . *P. jeffreyi*

P. albicaulis Engelm. Whitebark Pine. Low, often gnarled and twisted trees 1–10 m tall; bark thin, whitish to tan, smooth or broken by narrow fissures; lvs. in fives, 3.5–7 cm long, stout; cones ovoid to subglobose, 3.5–7 cm long, brown to brownish purple, remaining closed at maturity. Occasional. Exposed ridges at timberline, 8,000–9,000 ft. Subalpine F. This species was not collected but was observed on Caribou Mt. and Thompson Pk. July–August.

P. attenuata Lemmon. Knobcone Pine. Tree 3–12 m tall; bark on older trees thin, divided into low ridges with loose scales; lvs. in threes, 8–17 cm long, stiff; cones narrowly ovoid, asymmetrical, short-stalked, often whorled, the scales on the outer side enlarged into prominent knobs, remaining closed at maturity. Occasional. Dry slopes, 5,000–5,500 ft. Mixed Conifer F. Canyon Cr. (*717*); also observed at Big Flat and on Packers Pk. May.

P. balfouriana Grev. & Balf. in A. Murr. Foxtail Pine. Stout forest tree 3–25 m tall; bark thin, becoming thicker on older trees, light to red brown, divided into broad, flat ridges; lvs. in fives, 2.5–3.5 cm long, stout, densely covering the ± drooping ends of the branches; cones narrowly ovoid, 3.5–7.5 cm long, sessile, the scales with minute, incurved prickles. Common. High valleys, slopes, and ridges, 6,000–8,000 ft. Subalpine F. Packers Pk. (*561*); also observed at the head of Union Cr., Red Rock Mtn., Caribou Basin, and Grizzly L. Basin. July.

P. contorta Dougl. ssp. *latifolia* (Engelm. ex Wats.) Critchf. Lodgepole Pine. Trees with slender trunks 15–40 m tall, becoming gnarled, twisted, and ± prostrate at higher elevs.; bark thin, light brown to grayish, covered with small, thin scales; lvs. in twos, 3–6 cm long, slender; cones ovoid to subcylindric, often clustered, 2–5 cm long, dark brown, the scales armed with slender ± recurved prickles. Common. Dry slopes and flats, 5,000–7,500 ft. Mixed Conifer F., Red Fir F. Head of South Fork of the Salmon R. (*432*), Caribou Basin (*728*). July.

P. jeffreyi Grev. & Balf. in A. Murr. Jeffrey Pine. Forest tree 20–60 m tall, becoming stunted and somewhat gnarled at the upper limits of its range; bark on older trees thick, brown to reddish, separated into narrow to broad plates, often with a vanillalike odor; lvs. in threes, 12–28 cm long; cones broadly ovoid, 15–25 cm long on short stalks, the scales armed with downwardly deflexed prickles which are not noticed when the cone is grasped by the hand. Common. Slopes and ridges, valley floors at higher elevs., 5,500–8,000 ft. Subalpine F. This species was not collected but was observed throughout the study area at higher elevations. May–June.

P. lambertiana Dougl. Sugar Pine. Forest tree 20–75 m tall; bark on older trees 2.5–10 cm thick, reddish-brown, divided into elongate, rough ridges; lvs. in fives, 7–10 cm long, sharp pointed; cones cylindric, pendulous on long stalks on the uppermost branches, 25–45 cm long, light brown. Occasional. Valley floors and immediately adjacent slopes, 5,000–6,000 ft. Mixed Conifer F. This species was not collected but was observed at Big Flat and along the South Fork of the Salmon R. May–June.

P. monticola Dougl. Western White Pine. Forest tree 15–50 m tall, often forming pure stands; bark on older trees 2–4 cm thick, brown to gray, divided into small squarish plates; lvs. in fives, 5–10 cm long, obtuse; cones narrowly cylindric, pendulous on long stalks from near the ends of the uppermost branches, 10–20 cm long, light to dark brown. Common. Basins, slopes, and ridges, 6,000–7,000 ft. Red Fir F., Subalpine F. Caribou Basin (*189*). This species was observed throughout the study area at generally higher elevs.

P. ponderosa Dougl. ex P. & C. Lawson. Yellow Pine. Forest tree 15–70 m tall; bark on older trees 6–10 cm thick and separated into broad yellow-brown to reddish plates, with a strong resinous odor; lvs. in threes, 12–25 cm long; cones narrowly ovoid, 7–15 cm long on short stalks, the scales armed with outwardly protruding prickles which can be felt when grasped by the hand. Common. Bottoms of valleys and immediately adjacent slopes, 5,000–5,500 ft. Mixed Conifer F. This species was not collected but was observed throughout the study area at lower elevs. May–June.

Pseudotsuga Carr.

P. menziesii (Mirb.) Franco. Douglas-Fir. Forest tree to 70 m tall; bark becoming deeply fissured into broad ridges; lvs. 2–3 cm long; cones narrowly ovoid, 5–8 cm long, pendant; bracts well exserted from the scales. Common. Valley floors and immediately adjacent slopes, 5,000–6,000 ft. Mixed Conifer F. Packers Pk. (*565*); observed throughout the study area at lower elevs.

Tsuga Carr.

T. mertensiana (Bong.) Carr. Mountain Hemlock. Tree 20–45 m tall with a nodding leading shoot; bark deeply divided into rounded ridges covered with red-brown scales; branches horizontal to pendulous; lvs. 1.5–2 cm long, ± curved; cones oblong-cylindric, 2.5–7.5 cm long, sessile; bracts ca. one-quarter as long as the scales, sharp pointed at the tip, cuneate at the base.

Common. Exposed ridges and north-facing slopes, 6,500–8,500 ft. Subalpine F. Canyon Cr. (*183*); observed throughout the study area at higher elevs.

TAXACEAE

Taxus L.

T. brevifolia Nutt. Yew. Low, somewhat spreading tree 3–20 m tall; bark 5–6 mm thick, covered with small, reddish-brown scales; lvs. flat, 12–20 mm long, 1–2 mm wide, abruptly short pointed, forming flat sprays; fr. ovoid, 8–10 mm long. Occasional. Shaded canyons and slopes, 5,000 ft. Mixed Conifer F. Canyon Cr. (*781*); also observed at Big Flat. May.

DIVISION ANTHOPHYTA

Fls. mostly 4–5-merous; lvs. mostly net veined; cotyledons 2.
Dicotyledoneae p. 19

Fls. mostly 3-merous; lvs. mostly parallel veined; cotyledons 1.
Monocotyledoneae p. 145

Key to the families of the Dicotyledoneae

1. Petals absent, sepals sometimes petaloid.
  2. Plants parasitic on trees........................Loranthaceae p. 82
  2′. Plants not parasitic.
    3. Plants well developed shrubs or trees.
      4. Lvs. alternate.
        5. Calyx present in male fls.; fr. not a capsule.
          6. Fr. an acorn or spiny bur enclosing 1–3 nuts.
          Fagaceae p. 68
          6′. Fr. not an acorn or bur..............Betulaceae p. 24
        5′. Calyx absent; fr. a capsule................Salicaceae p. 118
      4′. Lvs. opposite ....Garryaceae p. 69
    3′. Plants herbaceous.
      7. Sepals lacking; the unisexual fls. borne in a cyathium.
      Euphorbiaceae p. 68
      7′. Sepals present; the fls. not borne in a cyathium.
        8. Pistils 3–many ....Ranunculaceae p. 104
        8′. Pistil 1.
          9. Sepals 4–6, less than 1 cm long ......Polygonaceae p. 91
          9′. Sepals 3, 3–5 cm long...........Aristolochiaceae p. 23
1′. Petals present.
  10. Petals separate. (See p. 21 for 10′.)
    11. Stamens more than twice the number of petals (See p. 20 for 11′.)
      12. Aquatic plants with floating lf. blades..Nymphaeaceae p. 82
      12′. Plants without floating lf. blades, mostly of drier places.
        13. Lvs. tubular, modified for trapping insects.
        Sarraceniaceae p. 118

13′. Lvs. not as above.
14. Lvs. with evident stipules.
15. Stamens united into a tube around the pistil. . . . . Malvaceae p. 82
15′. Stamens not so united. . . . . Rosaceae p. 110
14′. Lvs. without evident-stipules.
16. Plants well developed shrubs . . . . Rosaceae p. 110
16′. Plants herbaceous.
17. Lvs. opposite; pistil one. . . . . Hypericaceae p. 74
17′. Lvs. alternate; pistils many. . . . . . Ranunculaceae p. 104
11′. Stamens not more than twice the number of petals.
18. Plants well-developed trees or shrubs.
19. Ovary superior.
20. Fr. a samara. . . . . Aceraceae p. 22
20′. Fr. not a samara.
21. Stamens 10. . . . . (*Ledum*) Ericaceae p. 63
21′. Stamens 5–6.
22. Lvs. simple.
23. Stamens opposite petals. . . . . Rhamnaceae p. 109
23′. Stamens alternate with petals. . . . Celastraceae p. 34
22′. Lvs. pinnate. . . . . Berberidaceae p. 24
19′. Ovary inferior.
24. Fls. 4–merous; lvs. mostly opposite . . . . . . Cornaceae p. 54
24′. Fls. 5–merous; lvs. alternate . . (*Ribes*) Saxifragaceae p. 119
18′. Plants herbaceous.
25. Pistils 5 or more.
26. Lvs. succulent . . . . Crassulaceae p. 54
26′. Lvs. not succulent . . . . Rosaceae p. 110
25′. Pistil 1.
27. Styles 2 or more, separate to near base. (See p. 21 for 27′.)
28. Ovary superior.
29. Sepals 5.
30. Ovary with free central placentation. Caryophyllaceae p. 30
30′. Ovary with parietal or axial placentation. Saxifragaceae p. 119
29′. Sepals mostly 2; if more then plants have fleshy basal lvs. . . . . Portulacaceae p. 97
28′. Ovary at least half inferior.
31. Ovules solitary in each cavity of the locule; fls. in umbels . . . . Umbelliferae p. 139
31′. Ovules several in each cavity of the ovary; fls. not in umbels . . . . Saxifragaceae p. 119

27′. Style 1, sometimes ± divided toward apex.

32. Fls. irregular.

33. Anthers 5; fr. a capsule. . . . . Violaceae p. 143

33′. Anthers 10; fr. a legume . . . . Leguminosae p. 77

32′. Fls. regular.

34. Ovary inferior . . . . Onagraceae p. 83

34′. Ovary superior.

35. Sepals and petals 4; stamens mostly 6. Cruciferae p. 55

35′. Sepals and petals 5.

36. Stamens 5; fls. blue . . . . Linaceae p. 81

36′. Stamens 8–10; fls. not blue. . . . . Pyrolaceae p. 101

10′. Petals united, at least at base.

37. Ovary inferior.

38. Stamens 10 . . . . (*Vaccinium*) Ericaceae p. 63

38′. Stamens 5 or less.

39. Fls. in involucrate heads. . . . . Compositae p. 34

39′. Fls. not in involucrate heads.

40. Lvs. opposite or whorled.

41. Shrubs . . . . Caprifoliaceae p. 29

41′. Herbs.

42. Corolla 5-lobed; stamens 3 . . . . . . Valerianaceae p. 143

42′. Corolla 4-lobed; stamens mostly 4. . . . Rubiaceae p. 117

40′. Lvs. alternate . . . . Campanulaceae p. 27

37′. Ovary superior.

43. Fls. irregular. (See below for 43′.)

44. Plants lacking chlorophyll. . . . . Orobanchaceae p. 87

44′. Plants with green lvs.

45. Sepals 2; petals 4 in unlike pairs . . . . Fumariaceae p. 69

45′. Sepals 5; petals 3 or 5. . . . . (Go to 46.)

46. Petals 3, appearing to be 5 because 2 of the sepals are petaloid. Polygalaceae p. 91

46′. Petals 5, or apparently 4 by the fusion of the two upper ones; none of the sepals petaloid.

47. Stems round; fr. a capsule. . . . . Scrophulariaceae p. 127

47′. Stems square; fr. of 4 nutlets.

48. Ovary deeply 4-lobed, the style cleft or bifid. . . . . Labiatae p. 74

48′. Ovary not deeply lobed, the style entire. . . . . Verbenaceae p. 143

43′. Fls. regular.

49. Stamens 10 or more. (See p. 22 for 49′.)

50. Stamens many, filaments united. . . . . Malvaceae p. 82

50′. Stamens 10, filaments separate.

51. Pistils 5 . . . . Crassulaceae p. 54

51′. Pistil 1.
52. Plants lacking chlorophyll. (*Pterospora*) Pyrolaceae p. 101
52′. Plants with green lvs. . . . . Ericaceae p. 63
49′. Stamens 4, 5, or 6.
53. Pistils 2.
54. Petals and sepals strongly reflexed; corolla deeply lobed. Asclepiadaceae p. 23
54′. Petals and sepals erect; corolla not deeply lobed. Apocynaceae p. 23
53′. Pistil 1.
55. Stamens opposite corolla lobes . . . . . . Primulaceae p. 100
55′. Stamens alternate with corolla lobes. . . . . (Go to 56.)
56. Fls. small, dry-scarious; stamens 4. . . . . Plantaginaceae p. 87
56′. Fls. not dry-scarious, mostly colored; stamens 5.
57. Ovary 4-loculed, mostly 4-lobed, each lobe forming a nutlet in fr. Boraginaceae p. 25
57′. Ovary not 4-loculed; fr. a berry or capsule.
58. Style 3-cleft, at least toward the summit. . . . . . Polemoniaceae p. 88
58′. Style not 3-cleft.
59. Calyx toothed or cleft; style entire.
60. Lvs. opposite; corolla tubular or funnelform; fr. a capsule. Gentianaceae p. 71
60′. Lvs. alternate; corolla rotate; fr. a berry. . . . Solanaceae p. 139
59′. Calyx of 5 distinct sepals or sepals united only at base; style divided at summit.
61. Plants trailing; lvs. hastate; corolla plaited in bud. Convolvulaceae p. 53
61′. Plants not trailing; lvs. not hastate; corolla not plaited in bud. Hydrophyllaceae p. 71

ACERACEAE

Acer L.

Lvs. 10–25 cm wide, mostly 5-lobed; fls. in racemes . . *A. macrophyllum*
Lvs. 2–4 cm wide, mostly 3-lobed; fls. in corymbs. *A. glabrum* var. *torreyi*

A. glabrum Torr. var. *torreyi* (Greene) Smiley. Dwarf Maple. Fig. 1. Shrub or small tree, 2–6 m tall; twigs slender, reddish; lvs. deeply 3-lobed and doubly serrate, cordate at base, 2–4 cm wide, dark green above, dull green below, petioles slender, 1–5 cm long; fls. several in corymbs; sepals and petals greenish; samaras glabrous, 2–3 cm long, the wings spreading at 45 degrees. Occasional. Rocky canyon floors and slopes, 6,200 ft. Red Fir F. Kidd Cr. (*689*), Boulder Cr. Basin (*1293*). July.

A. macrophyllum Pursh. Big-leaved Maple. Tree to 30 m tall, bark gray, twigs reddish, coarse; lvs. 5-lobed, 10–25 cm wide, petioles 5–12 cm long; fls. numerous, in racemes 3–8 cm long; sepals and petals 3–4 mm long, greenish; samaras hairy, 2–4 cm long, the wings spreading. Occasional. Woods and canyons, 5,000–5,500 ft. Mixed Conifer F. Packers Pk. (*1145*). June.

APOCYNACEAE

Apocynum L.

A. pumilum (Gray) Greene. Mountain Dogbane. Perennial from a simple to branched woody caudex; stems 1-few, spreading to erect, with several branches, 15–30 cm tall, glabrous; lvs. opposite, spreading to drooping, glabrous; narrowly to broadly ovate, 1–4.5 cm long, dark green above, pale beneath, short petioled; fls. many in short cymes; calyx 1–2 mm long; corolla white to pinkish, 5–8 mm long; anthers connivent around the stigma; follicles erect at maturity, 4–12 cm long. Common. Dry, rocky slopes and meadows, 5,000–6,000 ft. Mixed Conifer F., Red Fir F. Big Flat (*253, 374*), near head of the South Fork of the Salmon R. (*408*). August.

var. *rhomboideum* (Greene) Bég. & Bel. Stems and at least the lower lvs. pubescent. Found at one location. Dry rocky slope, 6,000 ft. Red Fir F. Canyon Cr. (*807*). August.

ARISTOLOCHIACEAE

Asarum L.

A. caudatum Lindl. Long-tailed Wild Ginger. Perennial from slender, creeping rootstocks; lvs. basal with long, slender petioles, the blades cordate, pubescent, 3–8 cm broad; fls. solitary; calyx a deep, dark purple, paler in the throat, the lobes long-attenuate, 3–4 cm long; corolla absent; stamens 12, the anthers bearing short, terminal appendages; fr. a fleshy globular capsule. Found at one location. Stream bank, 5,000 ft. Mixed Conifer F. Grizzly Cr. (*929*). August.

ASCLEPIADACEAE

Asclepias L.

Fls. deep burgundy-purple; lvs. mostly glabrous and glaucous. *A. cordifolia*
Fls. pinkish or lavender; lvs. tomentose . . . . *A. speciosa*

A. cordifolia (Benth.) Jeps. Purple Milkweed. Fig. 2. Perennial from a stout, woody taproot; stems several, 2–4 dm tall, glabrous and somewhat glaucous; lvs. opposite with sessile, clasping bases, ovate, 2–7 cm long, glabrous and glaucous; fls. several in a loosely spreading umbel; sepals deep purple, ca. 2 mm long, white-woolly; petals deep burgundy-purple, strongly reflexed, ca. 4 mm long; hoods pale purplish, open at top and cleft down the inner surface, shorter than stamens, lacking a horn; follicles lanceolate, attenuate, glabrous, 4–6 (14) cm long. Occasional. Dry rocky slopes, 5,500–6,000 ft. Mixed Conifer F., Red Fir F. Packers Pk. (*556, 941*), Josephine L. Basin (*592*). June–August.

A. speciosa Torr. Showy Milkweed. Perennial from a ± woody caudex; stems 5–12 dm tall, glabrate to tomentulose; lvs. opposite, with short petioles, oval to ovate, 5–10 cm long, tomentose; fls. many in dense umbels, pedicels densely white tomentose; sepals purplish or lavender, tomentose, ca. 4 mm long; petals purplish to lavender, paler toward the base, strongly reflexed, 6–8 mm long; hoods pinkish aging yellow, much longer than the stamens, with an incurved horn exserted from the inner surface; follicles narrow-ovoid, densely woolly, 6–10 cm long. Found at one location. Dry meadow at edge of woods, 5,000 ft. Mixed Conifer F. Big Flat (*366*). July–August.

BERBERIDACEAE

Berberis L.

Leaflets 17–19; bud scales persistent .... *B. nervosa*
Leaflets mostly 7; bud scales deciduous.................. *B. piperiana*

B. nervosa Pursh. Oregon Grape. Shrubs 3–6 dm high; lvs. pinnately compound, tufted, 25–35 cm long; lfts. 17–19, oblanceolate, paler beneath, 3–7 cm long with 5–9 spine-tipped teeth along each side; fls. numerous, in erect to drooping racemes; sepals and petals yellowish-green, ca. 5 mm long; berries blue with a gray bloom, 8–10 mm long. Occasional. Woods, 5,000–6,000 ft. Mixed Conifer F. Packers Pk. (*1144*). June.

B. piperiana (Abrams) McMinn. Piper's Mahonia. Fig. 3. Low shrub to 4 dm tall; lvs. pinnately compound, 10–15 cm long, lfts. mostly 7, glossy green above, dull green beneath, 3–6 cm long with 7–11 spine-tipped teeth along each margin; fls. many in fascicled racemes; sepals 6, in 2 series, yellow, petal-like; petals 6, in 2 series, yellow; berry ellipsoid, blue-black, ca. 6 mm long. Found at one location. Wooded slope, 5,000 ft. Mixed Conifer F. Big Flat (*500*). June.

BETULACEAE

Fruit a nut enclosed in an involucre........................ *Corylus*
Fruit conelike .... *Alnus*

Alnus Hill.

A. sinuata (Regel) Rydb. Wavy-leaved Alder. Fig. 4. Shrub or small tree, 1–2 m tall; bark reddish brown to gray; lvs. broadly ovate, doubly serrate, dark green above, paler beneath, 3–7 cm long, the petioles 1–2.5 cm long; male and female fls. borne in catkins; male catkins 2.5–3 cm long; female catkins 10–14 mm long in fruit, forming a woody cone; nutlet oval, ca. 1.2 mm long, ca. as broad as wings. Occasional. Stream banks, 5,500–6,000 ft. Mixed Conifer F., Red Fir F. Caribou Basin (*192*), Kidd Cr. (*436*), Canyon Cr. (*775*). August.

Corylus L.

C. cornuta Marsh. var. *californica* (A. DC.) Sharp. California Hazelnut. Spreading shrub to 2 m high; bark grayish-brown; lvs. ovate to elliptic, irregularly serrate, 2–7 cm long, thinly pubescent with long hairs, dark green above, paler beneath, petioles short, stout, pubescent; male fls. borne in catkins; female fls. several from a scaly bud; fruit a smooth nut, 1.2–1.5 cm long, enclosed in a reddish, tubular, hispid involucre. Found at one location. Dry brushy slope, 5,000 ft. Red Fir F. Boulder Cr. Basin (*459*). August.

BORAGINACEAE

Nutlets bearing barbed spines; fls. showy.
  Nutlets evenly covered with spines . . . . *Cynoglossum*
  Nutlets bearing long spines on the margins and naked or with a few short spines on the back . . . . *Hackelia*
Nutlets not bearing spines; fls. inconspicuous.
  Nutlets attached by a thin ventral slit in the pericarp . . . . *Cryptantha*
  Nutlets lacking such a slit but bearing a ventral keel . . *Plagiobothrys*

Cryptantha Lehm.

C. affinis (Gray) Greene. Common Cryptantha. Widely branched annual; stems slender, thinly hispid, 15–25 cm tall; lvs. linear to oblanceolate, 1–3.5 cm long, hispid; fls. borne in several spikes; calyx 2–3 mm long, the linear lobes bearing long, stout hairs along the middle and shorter, finer hairs toward the margins; corolla white, ca. 1 mm broad; nutlets ca. 2 mm long, grayish and mottled, smooth, shiny, with an eccentric ventral slit. Found at one location. Steep wooded slope, 6,400 ft. Red Fir F. Kidd Cr. (*1107*). August.

Cynoglossum L.

C. occidentale Gray. Western Hound's Tongue. Fig. 5. Perennial from a stout taproot; stems hirsute, 1.5–2 dm tall; basal lvs. oblanceolate, hirsute, 3–10 cm long; gradually narrowed to winged petioles 4–8 cm long, upper lvs. sessile and ± clasping; infl. several fld., compact; calyx 5–6 mm long; corolla blue, prominently tinged brown to rose, 7–8 mm long, each lobe subtended by a scale; nutlets ca. 8 mm long covered with short barbed spines. Occasional. Dry open slopes, 5,000–6,000 ft. Mixed Conifer F., Red Fir F. Big Flat (*524*), Yellow Rose Mine Trail (*670*). June–July.

Hackelia Opiz

Corolla blue to pink, 4–6 mm broad . . . . *H. micrantha*
Corolla white, 10–20 mm broad . . . . *H. bella*

Acer glabrum var. torreyi

Asclepias cordifolia

Berberis piperiana

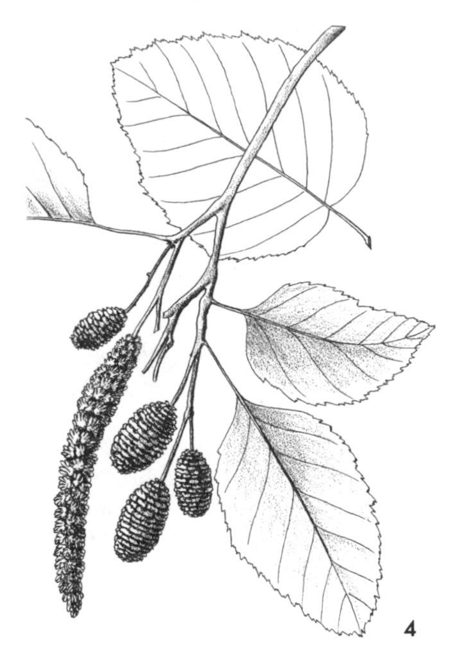

Alnus sinuata

H. bella (Macbr.) Jtn. Showy Stickseed. Fig. 6. Perennial from a branched caudex; stems few, 4–6 dm tall, pubescent; lvs. linear to lanceolate, 5–10 cm long, the basal on long petioles, the cauline sessile with clasping bases; fls. many in several racemes; pedicels stout and drooping in fruit; calyx 3–4 mm long; corolla white, 10–20 mm broad; nutlets 6–7 mm long, bearing long, slender, barbed bristles that are flattened at the base on the margins and a few very short bristles on the back. Occasional. Dry open or brushy slopes, 6,500 ft. Montane Chaparral, Red Fir F. Caribou Mt. (*211*), Dorleska Mine (*621*). July.

H. micrantha (Eastwood) J. L. Gentry. Jessica's Stickseed. Perennial from a branched caudex; stems usually several, 4–7 dm tall, hispid; lvs. oblanceolate, 5–13 cm long, the basal on long petioles, the cauline sessile and ± clasping; fls. many in several racemes; pedicels slender and drooping in fruit; calyx 2–3 mm long; corolla blue to pink, 4–6 mm broad; nutlets 4–5 mm long, bearing long barbed bristles that are flattened at the base on the margins and naked or with a few short bristles on the back. Common. Damp shaded places, 5,000–6,500 ft. Mixed Conifer F., Red Fir F. Along the South Fork of the Salmon R. (*91*), north side of Caribou Mt. (*212, 213*), Dorleska Mine (*622*), Caribou Basin (*737*). June–August. The nomenclature here follows Gentry (1972).

Plagiobothrys F. & M.

Nutlets keeled on back at apex, ± smooth; gynobase subulate. *P. cognatus*

Nutlets keeled on back to below middle, rugulose; gynobase pyramidal. *P. hispidulus*

P. cognatus (Greene) Jtn. Cognate Allocarya. Hispid annual; stems erect, simple to branched above, 10–20 cm tall; lvs. linear to narrowly oblanceolate, 10–35 mm long, 1–3 mm wide, the lower opposite, the upper alternate; calyx 4–5 mm long, the lobes linear; corolla white, ca. 2 mm broad; nutlets 2.2 mm long, shining, finely granular with a pronounced ventral keel and triangular basal scar. Found at one location. Open meadow, 5,000 ft. Mixed Conifer F. Big Flat (*605*). July.

P. hispidulus (Greene) Jtn. Harsh Allocarya. Low ascending to erect annual; stems mostly branched from base, 5–10 cm tall; lvs. linear, 10–25 mm long, ca. 1 mm wide; calyx 2–3 mm long, the lobes linear; corolla white, 1–1.5 mm broad; nutlets 1–1.5 mm long, shining, finely granular, the ventral keel not so pronounced, the scar narrow. Found at one location. Open meadow, 5,000 ft. Mixed Conifer F. Big Flat (*508*). June–July.

CAMPANULACEAE

Fls. regular; anthers separate.

Corollas ca. 3 mm long; capsule opening by irregular fissures. *Heterocodon*

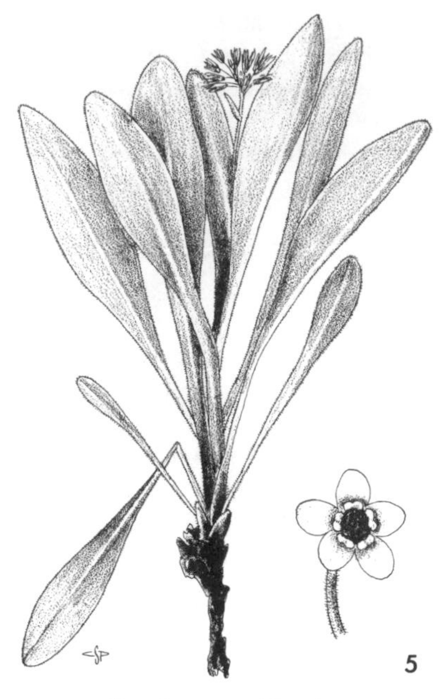

Cynoglossum occidentale

6

Hackelia bella

Campanula prenanthoides

Downingia yina

Corollas 7–15 mm long; capsule opening by lateral lids . *Campanula*
Fls. irregular; anthers united . . . . *Downingia*

Campanula L.

Style and stigma longer than and well exserted from the corolla.
C. prenanthoides
Style and stigma not longer than corolla . . . . *C. wilkinsiana*

C. prenanthoides Durand. California Harebell. Fig. 7. Perennial from slender rootstocks; stems slender, 2–6 dm tall, sparingly scabrous; lvs. ovate to lanceolate, sessile or short petioled, 1.5–5 cm long, coarsely serrate; fls. several; pedicels slender, 2–6 mm long; sepals subulate, 3–5 mm long; corolla blue, campanulate, 7–8 mm long with linear lobes; style and stigma 11–13 mm long, well exserted from the corolla. Common. Dry woods, 5,000 ft. Mixed Conifer F. Along the South Fork of the Salmon R. (*267*, *420*). July–August.

C. wilkinsiana Greene. Wilkins' Harebell. Perennial from slender rootstocks; stems slender, 0.5–2 dm tall; lvs. cuneate below, lanceolate above, mostly with a few slender teeth, sessile, 5–20 mm long, solitary at the ends of the stems; sepals subulate, 4–5 mm long; corolla blue, campanulate, 10–15 mm long, cleft ca. halfway into lanceolate lobes; style and stigma ca. as long as corolla. Occasional. Stream banks and damp meadows, 5,700–7,000 ft. Red Fir F. Kidd Cr. (*446*, *1099*), Canyon Cr. (*782*). August.

Downingia Torr.

D. yina Appleg. Cascade Downingia. Fig 8. Low annual; stems simple to branched from base, 3–10 cm high; lvs. linear, sessile, 4–7 mm long; fls. several, borne in the axils of leafy bracts, the inferior ovary appearing like a pedicel; sepals linear, 2–4 mm long; corolla blue, with yellow and white on the lower lip, 6–8 mm long; anther tube bearing 2 slender awns at the apex; capsule slender, 2–2.5 cm long. Occasional. Vernal ponds, 5,000–6,600 ft. Mixed Conifer F., Red Fir F. Dorleska Mine (*319*), Big Flat (*360*, *377*). August.

Heterocodon Nutt.

H. rariflorum Nutt. Heterocodon. Delicate, slender annuals, 7–25 cm high, simple to few branched; lvs. scattered, sessile, rounded, coarsely toothed, bristly-pubescent on margins, 3–6 mm long; fls. axillary, the corollas of the lower ones reduced and inconspicuous; sepals deltoid, 2–3 mm long; corolla of upper fls. blue, ca. 3 mm long; capsule ca. 2 mm in diam. Found at one location. Edge of drying seep, 5,200 ft. Mixed Conifer F. Boulder Cr. Basin (*1278*). July–August.

CAPRIFOLIACEAE

Fls. irregular, deep purple . . . . *Lonicera*

Fls. regular, white to cream . . . . *Sambucus*

Lonicera L.

L. conjugialis Kell. Double Honeysuckle. Fig. 9. Spreading shrub 5–20 dm tall; lvs. ovate, elliptic or obovate, 1–6 cm long, petioles 1–3 mm long; fls. borne in pairs with united ovaries on slender peduncles 1–2.5 cm long; calyx inconspicuous; corolla deep purple, 2-lipped, gibbous at base, 8–10 mm long; berry bright red, 5–6 mm long. Common. Stream banks and wooded slopes, 6,500 ft. Red Fir F. Caribou Basin (*191*), Kidd Cr. (*693*). July.

Sambucus L.

Lfts. 1.5–3 cm broad; fruit purple-black . . . . *S. melanocarpa*
Lfts. mostly 4–8 cm broad; fruit red . . . . *S. microbotrys*

S. melanocarpa Gray. Black Elderberry. Low shrub to 1 m tall; bark reddish brown; lvs. pinnate, ca. 10 cm long, lfts. mostly 5, lanceolate, 4–8 cm long, 1.5–3 cm broad, serrate, pubescent with stiff hairs; fls. many in a dense compound cyme 4–5 cm long; corolla white, turning brown on drying; fruit purplish-black, 4–5 mm in diam. Found at one location. Steep, damp, rocky slope, 7,000 ft. Subalpine F. Caribou Basin (*747*). July–August.

S. microbotrys Rydb. Mountain Red Elderberry. Shrub to 2 m tall; bark reddish brown; lvs. pinnate, 10–25 cm long, lfts. 5–7, obovate to elliptic, 2–12 cm long, 1–7 cm wide, coarsely serrate, glabrous to pubescent with a few hairs; fls. many in a dense compound cyme 3–6 cm long; corolla cream; fruit bright red, 5–6 mm in diam. Found at one location. Open slope, 6,500 ft. Red Fir F. Caribou Basin (*1080*). July-August.

CARYOPHYLLACEAE

Sepals united; petals clawed . . . . *Silene*
Sepals separate; petals not clawed.
  Styles mostly 3, sometimes 4; capsule ovoid, not bent.
    Petals entire.
      Stipules present, scarious . . . . *Spergularia*
      Stipules absent . . . . *Arenaria*
    Petals bifid at apex . . . . *Stellaria*
  Styles 5; capsule cylindrical, bent near summit . . . . *Cerastium*

Arenaria L.

Plants glabrous; fls. in dense heads . . . . *A. congesta*
Plants glandular-pubescent; fls. not in heads . . . . *A. nuttallii* ssp. *gregaria*

A. congesta Nutt. ex T. & G. Capitate Sandwort. Fig. 10. Glabrous perennial from a branched, woody caudex subtended by a horizontal rootstock; stems erect, 10–25 cm tall; basal lvs. crowded, filiform, 1–6 cm long, cauline lvs.

somewhat reduced, in 3–4 remote pairs; infl. many fld., dense, ± capitate; sepals with a green midrib and scarious margins, 3–4 mm long; petals white, 4–6 mm long; capsules exserted from calyx. Common. Dry, rocky slopes, 5,600–7,500 ft. Red Fir F., Subalpine F. Dorleska Summit (*121*), Canyon Cr. (*163, 779*), Caribou Basin (*199*), north side of Caribou Mt. (*330*), Packers Pk. (*932*). June–August.

A. nuttallii Pax ssp. *gregaria* (Heller) Maguire. Nuttall's Sandwort. Aromatic, glandular-pubescent perennial from a slender, branched caudex; stems diffusely branched, procumbent to ascending, 5–15 cm high; lvs. narrowly subulate, crowded, 5–7 mm long; fls. several in open cymes; pedicels slender, 5–12 mm long; sepals mostly green with narrow hyaline margins, 3–4 mm long; petals white, 3–4 mm long; capsule included in calyx. Occasional. Loose, rocky slopes, mostly serpentine, 7,000–8,000 ft. Red Fir F., Subalpine F. Dorleska Summit (*304*), Red Rock Mt. (*974, 1022*). August.

Cerastium L.

Petals twice as long as sepals .... *C. arvense*
Petals ca. equal to sepals .... *C. vulgatum*

C. arvense L. Field Chickweed. Pubescent perennial; stems 1–several, 10–20 cm tall; lvs. linear to lanceolate, 8–15 mm long; fls. few in an open cyme; sepals 4–5 mm long with hyaline margins; petals white, 8–10 mm long; capsule ca. equal to calyx. Occasional. Open meadows, 5,000–6,000 ft. Mixed Conifer F., Red Fir F. Along the South Fork of the Salmon R. (*96*), Bullards Basin (*630*). June–July.

C. vulgatum L. Larger Mouse-ear Chickweed. Pubescent perennial from slender rootstocks; stems 10–15 cm tall; lvs. lanceolate, oblong, or oblanceolate, 5–15 mm long; fls. few in open cymes; sepals 5–6 mm long, with purplish tips and hyaline margins; petals white, ca. equal to sepals; capsule longer than calyx and exserted from it. Found at one location. Hillside seep, 5,000 ft. Mixed Conifer F. Big Flat (*491, 529*). June–July.

Silene L.

Cauline lvs. well developed .... *S. douglasii*
Cauline lvs. reduced.
    Styles and stamens greatly exserted; claw of petals pubescent; fls. nodding .... *S. lemmonii*
    Styles and stamens not so exserted; claw of petals glabrous; fls. usually erect .... *S. grayi*

S. douglasii Hook. Douglas' Campion. Perennial from a branched, woody caudex; stems few–several, glabrous to finely strigulose, 25–40 cm tall; basal lvs. not well developed, cauline lvs. linear-lanceolate to oblanceolate, 1–5 cm

Lonicera conjugialis

Arenaria congesta

Silene grayi

Stellaria longipes

long, glabrous to scabrous; fls. few, erect; calyx 12–15 mm long, strigulose; petals white sometimes tinged pink or purple, ca. 20 mm long, 2-cleft, appendages 2, ca. 2 mm long, claws glabrous; styles and stamens not much exserted. Occasional. Open to wooded, rocky slopes, 7,000–7,500 ft. Red Fir F. Grizzly L. Basin (*887*), Sawtooth Ridge (*1069*). August.

S. *grayi* Wats. Gray's Campion. Fig. 11. Perennial from a branched, woody caudex; stems few–several, strigulose, becoming glandular above; basal lvs. tufted, oblanceolate, 1–2.5 cm long, strigulose, cauline lvs. in 2–3 remote pairs, reduced; fls. several, erect; calyx ca. 10 mm long, glandular-pubescent; petals white to rose-tinted, ca. 15 mm long, 4-lobed, the 2 lateral lobes ca. half as long as the 2 middle lobes, appendages 2, ca. 1 mm long; claws glabrous; styles and stamens not much exserted. Common. Open rocky ridges and slopes, 7,000–7,500 ft. Red Fir F., Subalpine F. Dorleska Summit (*607*), Caribou Basin (*753*), Red Rock Mt. (*1030*). July–August.

S. *lemmonii* Wats. Lemmon's Campion. Perennial from a branched, woody caudex; stems few, pubescent, becoming glandular above, ca. 15 cm tall; basal lvs. oblanceolate, 1–2.5 cm long, densely stiff-pubescent; cauline lvs. few and greatly reduced; fls. few, nodding at anthesis; calyx 8–10 mm long, glandular-pubescent; petals white, ca. 12 mm long, divided into 2 lobes which are again 2-cleft, appendages 2, ca. 1 mm long, claw hairy; styles and stamens greatly exserted. Found at one location. Dry wooded slope, 5,200 ft. Mixed Conifer F. Packers Pk. (*554*). June–July.

Spergularia J. & C. Presl.

S. *rubra* (L.) J. & C. Presl. Purple Sand Spurry. Prostrate perennial; stems ± matted, 5–10 cm long, glabrous below infl., lvs. opposite, fascicled, acicular, 5–10 mm long; stipules acuminate, hyaline; infl. sparingly to densely glandular-pubescent; sepals ca. 3 mm long, with hyaline margins; petals pink, ca. 2 mm long; capsule as long as sepals; pedicels slender, spreading in fr., 7–8 mm long. Occasional. Disturbed places, 5,000 ft. Mixed Conifer F. Big Flat (*1126, 1194*). June.

Stellaria L.

Petals shorter than sepals, ca. 1.5 mm long . . . . S. *crispa*
Petals equal to or longer than the sepals, 5–8 mm long.
    Lvs. linear, 6–20 mm long; plants glabrous . . . . S. *longipes*
    Lvs. lanceolate, 25–40 mm long; plants glandular-pubescent, at least above . . . . S. *jamesiana*

S. *crispa* Cham. & Schlecht. Chamisso's Starwort. Glabrous perennial from slender rootstocks; stems weak, simple, 10–15 cm tall; lvs. thin, ovate, 5–17 mm long; fls. few; pedicels axillary, slender, 6–20 mm long; sepals 2.5–3 mm long with scarious margins; petals white, ca. 1.5 mm long, deeply bifid; styles mostly 4. Found at one location. Stream bank, 7,300 ft. Red Fir F. Caribou Basin (*1058*). August.

S. *jamesiana* Torr. Sticky Starwort. Perennial from slender rootstocks; stems 15–25 cm tall, glandular-pubescent, at least above; lvs. lanceolate, 25–40 mm long, glabrous to glandular-pubescent; fls. several in an open panicle; sepals **4–5 mm long, glandular-pubescent**; petals white, 5–8 mm long, bifid; styles 2. Found at one location. Dry opening in woods, 5,000 ft. Mixed Conifer F. Big Flat (*535*). June–July.

S. *longipes* Goldie. Long-stalked Starwort. Fig. 12. Glabrous perennial from slender rootstocks; stems slender, 10–25 cm tall; lvs. linear-lanceolate, 6–20 mm long; fls. 1–few; sepals ca. 4 mm long; petals white, ca. 5 mm long, bifid; styles 3. Found at one location. Boggy meadow, 5,000 ft. Mixed Conifer F. Big Flat (*140*). June–July.

CELASTRACEAE

Paxistima Raf.

P. *myrsinites* (Pursh) Raf. Oregon Boxwood. Low, spreading shrub 3–6 dm high; lvs. opposite, elliptic to obovate, serrulate toward apex, shiny above, 1–3 cm long on short petioles; fls. small, petals ca. 1 mm long, reddish brown; capsule 4–5 mm long. Occasional. Woods, 5,000–6,000 ft. Mixed Conifer F. Big Flat (*1132*). June–July.

COMPOSITAE

Plants thistlelike .... *Cirsium*
Plants not thistlelike.
  Ray fls. absent.
    Pappus absent.
      Stems leafy only at base, glandular above . . *Adenocaulon*
      Stems leafy throughout, not glandular. . . . . *Artemisia*
    Pappus present.
      Pappus of paleaceous scales .... *Chaenactis*
      Pappus of capillary bristles.
        At least some lvs. opposite.
          Phyllaries strongly nerved. . . . . *Brickellia*
          Phyllaries not nerved .... *Arnica*
        Lvs. all alternate or basal.
          Phyllaries completely scarious or hyaline, white to pinkish or sordid.
            Plants from woody stolons or rhizomes, often mat forming.
              Cauline lvs. mostly reduced upward, basal lvs. prominent; phyllaries most-

ly sordid to whitish or pinkish.
Antennaria
Cauline lvs. not much reduced, basal lvs. early deciduous; phyllaries pearly white . . . . *Anaphalis*
Plants from slender, delicate, fibrous roots, not mat forming . . . . *Gnaphalium*
Phyllaries not completely scarious or hyaline, mostly with some green, herbaceous to chartaceous.
Lvs. mostly entire . . . . *Erigeron*
Lvs. sharply serrate . . . . *Haplopappus*
Ray fls. present.
Ray fls. making up the entire head.
Corollas pink to purple . . . . *Stephanomeria*
Corollas white, yellow, or orange.
Pappus plumose . . . . *Microseris*
Pappus barbed but not plumose.
Corollas orange . . . . *Agoseris*
Corollas bright yellow or white.
Lvs. entire . . . . *Hieracium*
Lvs. ± divided or lobed.
Heads few to many . . . . *Crepis*
Heads solitary . . . . *Taraxacum*
Ray fls. marginal only, the inner fls. tubular.
Pappus absent.
Corollas white . . . . *Achillea*
Corollas yellow . . . . *Madia*
Pappus present.
Pappus of paleaceous scales.
Plants densely white-woolly . . . . . . *Eriophyllum*
Plants not densely white-woolly.
At least one of the pappus scales on each achene bearing a stout awn; plants of dry places . . . . *Wyethia*
All of the pappus scales attenuate into a slender awn; plants of wet places.
Helenium
Pappus of slender bristles.
Rays blue or purple.
Phyllaries mostly in 2 or more series, sometimes indistinct, mostly imbricate . . . . *Aster*
Phyllaries mostly in one series, mostly not imbricate . . . . *Erigeron*
Rays yellow or orange.

Rays orange .... *Raillardella*
Rays yellow.
At least some lvs. opposite.... *Arnica*
Lvs. all alternate........... Go to A

A. Plants low, rounded shrubs or cespitose herbs...... *Haplopappus*
AA. Plants not shrubs or cespitose.
Heads small, invol. ca. 3 mm high, phyllaries ± imbricate. *Solidago*
Heads larger, invol. 4–10 mm high, phyllaries in one series. *Senecio*

Achillea L.

A. lanulosa Nutt. Yarrow. Fig 13. Aromatic perennial from a slender rhizome, the entire plant ± villous-pubescent; stems simple, 2–7 dm tall; lvs. 4–15 cm long, mostly less than 1.5 cm wide, pinnately dissected into narrow, ultimate divisions; heads numerous; invol. ca. 4 mm high, phyllaries imbricate, with brownish margins; rays few, broad, white; disk fls. ca. 20; achenes 2 mm long. Common. Dry, open slopes, 5,000–8,100 ft. Mixed Conifer F., Red Fir F., Subalpine F. Caribou Basin (*196*), Big Flat (*247*), Caribou Mt. (*334*), Red Rock Mt. (*1028*). July–August.

Adenocaulon Hook.

A. bicolor Hook. Trail-plant. Fig. 14. Perennial from slender rootstocks; stems slender, 2–5 dm tall, floccose-woolly below, glandular above; lvs. deltoid-ovate, ± cordate at base, 4–7 cm long on petioles ca. as long, glabrous above, densely white-woolly beneath, coarsely sinuate-dentate; branches of infl. slender and glandular; phyllaries 1.5–2 mm long, reflexed in fruit and then deciduous; heads few fld.; achenes 5–8 mm long, glandular above. Common. Shaded woods, 5,000 ft. Mixed Conifer F. Along South Fork of Salmon R. (*264, 405*), Canyon Cr. (*846*). August.

Agoseris Raf.

A. aurantiaca (Hook.) Greene. Orange-flowered Agoseris. Scapose perennial from a taproot and with milky juice; scapes few to several, 1–5 dm tall, loosely villous becoming densely so toward the heads; lvs. linear-lanceolate, variously toothed or lobed, loosely villous, 5–20 cm long; inner phyllaries lanceolate, often purple spotted, 10–20 mm long, outer phyllaries half as long; rays burnt orange, turning purple in age; achenes narrow-fusiform, 4–8 mm long, strongly ribbed, tapering into a slender beak at least 2–4 mm long; pappus white, of many scabrid, capillary bristles. Found at one location. Open meadow, 6,300 ft. Red Fir F. Caribou Mt. (*322*). August.

Anaphalis DC.

A. margaritacea (L.) Benth. ex C. B. Clarke. Pearly Everlasting. Perennial from creeping rootstocks; stems 3–6 dm high; cauline lvs. numerous, lanceolate with clasping bases, 3–8 cm long, gray-tomentose below, tomentose to glabrate above; heads numerous; invol. woolly at base, 4–5 mm high, phyllaries ovate, pearly white, often with a black or pink area at base; pappus bristles translucent, often barbed upward, appearing jointed; achenes ca. 1 mm long. Occasional. Dry rocky places, 5,000–6,000 ft. Mixed Conifer F., Red Fir F. Boulder Cr. Basin (*1283*), Canyon Cr. (*Hitchcock & Martin 5361*). July–August.

Antennaria Gaertn.

Phyllaries white . . . . *A. argentea*
Phyllaries colored.
  Phyllaries dirty blackish-green . . . . *A. alpina* var. *media*
  Phyllaries pink to rose . . . . *A. rosea*

A. alpina (L.) Gaertn. var. *media* (Greene) Jeps. Alpine Everlasting. Fig. 15. Dioecious, stoloniferous, mat-forming perennial; flowering stems erect to ascending, 4–15 cm tall, densely white-tomentose; lvs. of basal shoots narrowly oblanceolate to spatulate, 1–2 cm long, tomentose to glabrate, cauline lvs. fewer and narrower; female phyllaries white-woolly below the blackish-green tips; achenes glandular; female pappus dense, of capillary bristles, male pappus bristles dilated at tips. Common. Open, damp, rocky slopes and meadows, 7,000–8,000 ft. Subalpine F., Alpine Fell-fields. North side of Thompson Pk. (*920*), Caribou Basin (*1048*). August.

A. argentea Benth. Silvery Everlasting. Dioecious rhizomatous perennial; stems 2–4 dm tall, loosely tomentose; lvs. mostly basal, oblanceolate to spatulate, 2–4 cm long, cauline lvs. fewer, reduced upwards, ± tomentose; phyllaries 2–3 mm long, white, glabrous; achenes finely puberulous; female pappus bristles linear, united at base, male pappus bristles dilated and ± serrulate at tips. Occasional. Open woods, 5,000 ft. Mixed Conifer F. Along South Fork of Salmon R. (*265*). July–August.

A. rosea Greene. Rosy Everlasting. Dioecious perennial forming leafy mats; stems 1–3 dm tall, tomentose, with scattered narrow lvs.; basal lvs. narrowly oblanceolate to spatulate, 1–3 cm long, tomentose; heads numerous, in dense capitate clusters; phyllaries 2–4 mm long, white-woolly below, pinkish to roseate tipped; achenes glabrous, female pappus of barbellate, capillary bristles. Occasional. Open, wooded slopes, 5,500–6,000 ft. Mixed Conifer F., Red Fir F. Josephine L. Basin (*80*), Grizzly Cr. (*868*). June–August.

Arnica L.

Heads discoid.
  Lower lvs. petiolate; heads erect . . . . *A. discoidea* var. *eradiata*
  All lvs. sessile; heads pendent . . . . *A. viscosa*
Heads radiate.

Cauline lvs. in 5 or more pairs .... *A. amplexicaulis*
Cauline lvs. in 2–4 pairs.
Pappus tawny, subplumose.
Lower lvs. the largest .... *A. mollis*
Middle lvs. the largest .... *A. diversifolia*
Pappus whitish, barbellate.
Achenes glabrate to glabrous below the middle. *A. latifolia*
Achenes short-hairy and ± glandular to the base. *A. cordifolia*

A. *amplexicaulis* Nutt. Streambank Arnica. Perennial from a thickened rhizome; stems 1–2 dm tall, glabrous; lvs. in 5 pairs, narrowly elliptic, glabrous except for stiff hairs along margin, sinuately dentate, the lower short petioled the upper sessile and ± clasping, the middle pairs the largest; heads 3–4; phyllaries ca. 10 mm long, sparingly hirsute and glandular; rays bright yellow, 10–20 mm long; achenes dark, hirsute; pappus tawny, barbellate to subplumose. Rare. In cracks in granite cliffs where seeps occur, 5,800 ft. Red Fir F. Boulder Cr. Basin (*464*). August. This species has been previously reported for Calif. only from the Sierra Nevada (Munz, 1959).

A. *cordifolia* Hook. Heart-leaved Arnica. Perennial from slender rhizomes; stems 1–3 dm tall, mostly single, loosely white-hairy and glandular-puberulent; cauline lvs. in ca. 3 pairs, lanceolate to round-ovate, the lower petiolate the upper sessile, sinuately dentate, glandular and with some long hairs, the lower lvs. cordate to truncate at base; heads 1–3; phyllaries mostly in 1 series, 10–20 mm long, pilose below, glandular-puberulent above; rays 15–30 mm long, yellow; achenes dark brown, short hairy and ± glandular; pappus whitish, long-barbellate. Common. Open woods, 6,000–8,000 ft. Red Fir F., Subalpine F. Canyon Cr. (*159*), Caribou Mt. (*338*). June–August.

A. *discoidea* Benth. var. *eradiata* (Gray) Cronq. Rayless Arnica. Perennial from a slender rhizome; stems mostly single, often branched above, 2–4 dm tall, glandular-puberulent and with longer hairs; lvs. mostly in 3 pairs, ovate to deltoid, the lower wing-petioled, often cordate at base, the upper sessile, pubescent with white ± shaggy hairs, dentate to ± entire; heads 2–5; phyllaries 10–15 mm long, hirsute below and on margins above, ± glandular throughout; corollas all tubular, ± hirsute especially on lobes; achenes brown to gray, pubescent with stiff hairs and ± glandular; pappus whitish, barbellate. Found at one location. Wooded, rocky slope, 5,000 ft. Mixed Conifer F. Canyon Cr. (*63*). June.

A. *diversifolia* Greene. Lawless Arnica. Perennial from freely rooting rhizomes; stems solitary, 2–2.5 dm tall, glandular-puberulent and with scattered longer hairs; lvs. in 3 pairs, ovate, glandular-puberulent, irregularly dentate, the lower petioled the upper sessile; heads 1–5; phyllaries 7–10 mm long, glandular and ± hirsute; rays 10–15 mm long, yellow, disk corollas glabrous on back of lobes; achenes brown, pubescent with stiff hairs; pappus tawny, subplumose. Rare. Wet cracks in granite walls, 6,200 ft. Red Fir F. Boulder Cr. Basin (*479*). August. This species has been previously reported for Calif. only from the Sierra Nevada (Munz, 1959).

A. latifolia Bong. Mountain Arnica. Perennial from slender, brownish, naked rhizomes; stems 2–3 dm tall, glandular-puberulent and with longer hairs, often branched above; cauline lvs. in 3 pairs, mostly sessile, sparingly pilose and glandular, deltoid to lance-ovate, sinuately dentate, the basal lvs. petioled; heads 3–5; phyllaries 10–13 mm long, pilose at base and along upper margins, ± glandular; rays 10–15 mm long, yellow, disk corollas pilose on back of lobes; achenes glabrate below, sparingly hairy and glandular above; pappus whitish, barbellate. Occasional. Damp, shaded woods, 6,000–7,500 ft. Red Fir F. Browns Meadow (*765*), Caribou Basin (*1042*). July–August.

A. mollis Hook. Cordilleran Arnica. Fig. 16. Perennial from thickish, freely rooting rhizomes; stems ca. 3 dm tall, branched above, sparingly pilose and glandular, peduncles becoming densely so below the heads; basal lvs. lance-ovate, petioled, sparingly glandular-pilose, cauline lvs. sessile in 2–3 pairs, lance-ovate, glandular-pilose; heads ca. 4; phyllaries ca. 10 mm long, glandular-pilose; rays 10–15 mm long, yellow; some disk corollas with a few hairs on back of lobes; achenes ± pubescent throughout; pappus tawny, subplumose. Common. Stream banks and meadows, 6,000–7,800 ft. Red Fir F. Ward L. (*1087*) Canyon Cr. (*1295*, *1298*), Boulder Cr. Basin (*1422*). August. This species has been previously reported for Calif. only from the Sierra Nevada (Munz, 1959).

A. viscosa Gray. Shasta Arnica. Fig. 17. Stout perennial, forming dense clumps; stems many, angled, 4–5 dm tall, ± glandular-pilose, with numerous slender branches; lvs. sessile, alternate to opposite, glandular-pilose; heads solitary on each branch, pendent; phyllaries 5–10 mm long, ± glandular-pilose; corollas greenish-white, glabrous on back of lobes; achenes brown, thinly glandular-pilose; pappus somewhat tawny, strongly barbellate. Rare. Open, rocky meadow, 6,200 ft. Red Fir F. Boulder Cr. Basin (*484*). August. This species has been previously reported for Calif. only from Mt. Shasta (Munz, 1959).

Artemisia L.

Lvs. few-lobed or coarsely toothed, densely gray-tomentose on lower surface . . . . *A. douglasiana*
Lvs. deeply dissected, mostly glabrous . . . . . . *A. norvegica* var. *saxatilis*

A. douglasiana Bess. in Hook. Douglas' Mugwort. Rhizomatous perennial; stems 6–8 dm high, pubescent; lvs. numerous, 5–11 cm long, oblanceolate to

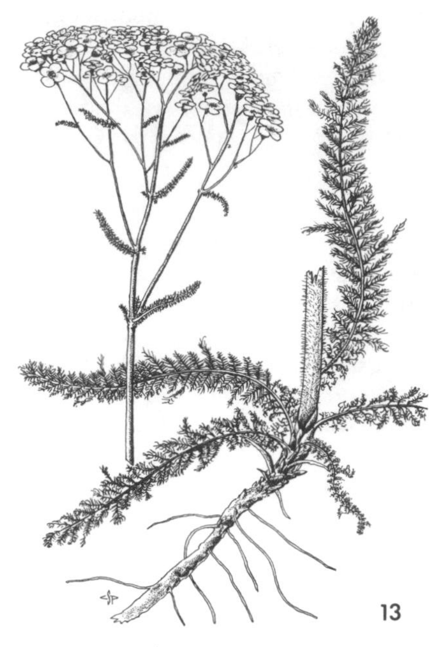

Achillea lanulosa

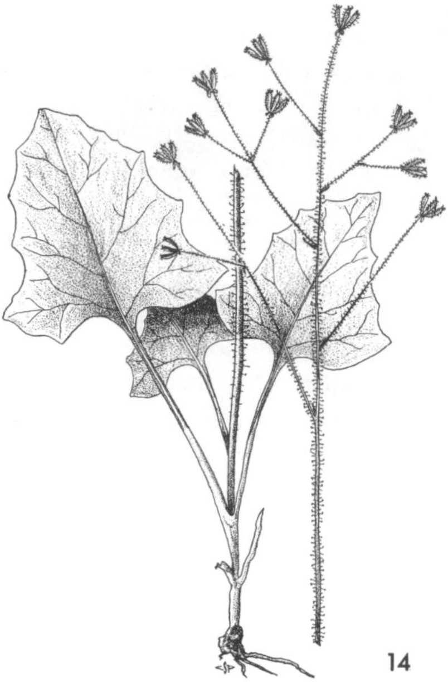

14

Adenocaulon bicolor

Antennaria alpina var. media

Arnica mollis

elliptic in outline, coarsely few toothed or lobed, tapering to a narrow tip, dull green and sparingly pubescent above, densely gray-tomentose beneath; heads numerous, small; invol. ca. 3 mm high, phyllaries tomentose with hyaline margins; corollas yellowish, glabrous; achenes ca. 1 mm long. Found at one location. Dry slope, 5,000 ft. Mixed Conifer F. Big Flat (*1371*). August–September.

A. norvegica Fries. var. *saxatilis* (Bess. in Hook.) Jeps. Mountain Sagewort. Fig. 18. Stout, glabrous perennial from a heavy, branched caudex; stems 3–4 dm tall, in dense clumps; lvs. ovate in outline, greatly dissected, the ultimate segments narrow and acute; heads numerous, pendulous; phyllaries 3–4 mm long, ± imbricate in 2 series, the margins dark brown; corollas with a few hairs at very base, yellowish green; achenes light brown, 1–1.5 mm long. Rare. Open rocky slope, 8,000 ft. Alpine Fell-field. Caribou Mt. (*342*). August. This species has been previously reported for Calif. only from the Sierra Nevada and the White Mts. (Munz, 1959).

Aster L.

Lower lvs. well developed, prominent.
  Heads solitary; plants from a taproot. . *A. alpigenous* ssp. *andersonii*
  Heads mostly 2–6; plants from creeping rhizomes . . *A. occidentalis*
Lower lvs. reduced and scale-like or deciduous at flowering.
  Phyllaries glandular, purple at tips and along margins.
    A. ledophyllus
  Phyllaries not glandular, entirely green . . . . *A. eatonii*

A. alpigenus (T. & G.) Gray ssp. *andersonii* (Gray) Onno. Alpine Aster. Subscapose perennial from a stout caudex surmounting a fleshy taproot; stems 1.5–3 dm tall, glabrous below becoming densely tomentose toward the heads; basal lvs. lance-linear, tufted, 5–25 cm long, 2–13 mm wide, cauline lvs. few, sessile, greatly reduced; heads solitary; phyllaries 8–15 mm long, ± tomentose, the inner with purple margins; rays numerous, purple, disk corollas yellow; achenes densely pilose; pappus of many capillary bristles, ± tawny. Occasional. Open damp meadows, 6,700–7,100 ft. Red Fir F. Ward L. (*451*), ridge west of Bullards Basin (*967*). July–August.

A. eatonii (Gray) Howell. Eaton's Aster. Stems from creeping rhizomes, 4–6 dm high, sparingly pubescent below, becoming densely so toward heads; lvs. numerous, linear-lanceolate, 3–10 cm long, tapering to sessile bases, lower lvs. early deciduous; heads few to numerous crowded toward ends of branches, subtended by foliaceous bracts; phyllaries scarcely graduate, green, ciliate, oblanceolate, 3–5 mm long; pappus tawny. Found at one location. Edge of bog, 5,000 ft. Mixed Conifer F. Big Flat (*1372*). August–September.

A. ledophyllus (Gray) Gray. Cascade Aster. Fig. 19. Perennial from a

17

Arnica viscosa

18

Artemisia norvegica var. saxatilis

19

Aster ledophyllus

20

Brickellia greenei

woody, often branched caudex; stems usually several, 2–6 dm tall, glabrous; lowermost lvs. reduced to brownish scales, the others numerous and uniform, narrowly lanceolate to elliptic-oblong, 1–4 cm long, sessile, glabrate above, ± tomentose below; heads several; rays few, purple, disk corollas yellow; invol. 7–10 mm high, phyllaries 4–5 seriate, ± glandular and ciliate, purple at the tips and along the upper margins; achenes loosely pilose; pappus of numerous, capillary, minutely barbellate bristles. Common. Dry, open, rocky meadows and slopes, 5,700–7,500 ft. Red Fir F., Subalpine F. Josephine L. (*294*), Yellow Rose Mine (*298*), Canyon Cr. (*804*), Grizzly Cr. (*873*), Red Rock Mt. (*977, 1031*), head of Union Cr. (*987*), Kidd Cr. (*1108*). August.

A. occidentalis (Nutt.) T. & G. Western Mountain Aster. Stems from creeping rhizomes, 1.5–3 dm high, glabrous below thinly pubescent above; lvs. linear-oblanceolate, 3–8 cm long, the basal tapering to narrow-winged, ciliate petioles; cauline lvs. tapering to sessile bases, becoming reduced upwards; heads solitary to few, at the ends of the branches; phyllaries in 2 indistinct series, linear to oblanceolate, 3–5 mm long, ± pubescent, green; pappus tawny; achenes thinly pubescent. Found at one location. Edge of dry lake, 5,000 ft. Mixed Conifer F. Big Flat (*384*). July–August.

Brickellia Ell.

Heads clustered; lvs. 2–4 cm long. *B. grandiflora*
Heads solitary; lvs. 1.5–2 cm long . . . . *B. greenei*

B. grandiflora (Hook.) Nutt. Large-flowered Brickellia. Perennial from a woody, branched caudex; stems slender, puberulent, 1–3 dm tall; lvs. usually opposite, the blades deltoid-ovate to lanceolate, 2–4 cm long, dentate, puberulent above, glandular-punctate below, the slender petioles 5–15 mm long; heads numerous, clustered and ± drooping at the ends of paniculately arranged branchlets; phyllaries lanceolate, loosely imbricate, ciliate, strongly nerved; corollas greenish white; achenes pilose above; pappus of many capillary, subplumose bristles. Found at one location. Talus slope, 7,000 ft. Subalpine F. Ridge between Landers L. and Union L. (*979*). August.

B. greenei Gray. Greene's Brickellia. Fig. 20. Perennial from a woody, branched caudex; stems many, slender, 1–2 dm tall, glandular-pilose; lvs. mostly alternate, ovate, serrate, glandular-pubescent, 1.5–2 cm long on petioles 1–2 mm long; heads solitary; phyllaries linear, glabrous, purple at the tips, strongly nerved; corollas greenish-white to pale yellow; achenes with a few hairs at the very top; pappus of many capillary, barbellate bristles with purple spots toward the tips. Found at one location. Open, rocky slope, 5,400 ft. Montane Chaparral. Canyon Cr. (*842*). August.

Chaenactis DC.

C. douglasii (Hook.) H. & A. Hoary Chaenactis. Low herbs from an elongate taproot; stems ca. 1 dm tall, ± tomentose; lvs. mostly basal, pinnatifid, glandular-punctate and ± tomentose, the blades 2–4 cm long; heads few; phyllaries narrowly oblanceolate, ± glandular; corolla whitish; achenes finely pubescent, 7–8 mm long; pappus of ca. 10 linear-oblong paleae. Occasional. Open, dry, gravelly slopes, 7,500 ft. Red Fir F. Black Mt. (*1086*). August.

Cirsium Mill.

C. callilepis (Greene) Jeps. Fringe-bract Thistle. Stout perennial 3–12 dm tall, the stems ± arachnoid; lvs. narrow, 0.5–3 dm long, green and glabrate above, paler and thinly tomentose beneath, ± spinose, sinuately toothed to deeply pinnatifid, the upper auriculate-clasping; heads numerous, ± rounded; phyllaries closely imbricate, 3–4 seriate, laciniate-fringed toward the tips, the outer and middle with a stout spine at the tip; corolla cream; achenes light brown with darker markings, oblong, ca. 6 mm long; pappus somewhat tawny, of numerous, finely plumose bristles. Occasional. Open woods, 5,000 ft. Mixed Conifer F. Big Flat (*263*). July–September.

Crepis L.

C. pleurocarpa Gray. Naked-stemmed Hawksbeard. Perennial from a slender taproot; stems 1.5–4 dm tall, ± tomentulose with milky juice; basal lvs. 10–15 cm long, glabrate to tomentulose, narrowly elliptic in outline, denticulate to pinnately divided with deltoid to lanceolate lobes; heads few to many, narrow, 4–12 fld.; outermost phyllaries few and short, the others 8–13 mm long, greenish, thinly tomentulose in the middle, becoming densely so toward the scarious margin; rays yellow; achenes dark brown, 10-ribbed, narrowed to a very short beak; pappus white, of numerous capillary, minutely barbed bristles. Occasional. Dry rocky slopes and meadows, 6,500–7,750 ft. Red Fir F. Yellow Rose Mine (*300*), head of Union Cr. (*993*). August.

Erigeron L.

Heads discoid.
  Middle and upper part of stems spreading villous. . . . *E. petrophilus*
  Middle and upper part of stems glabrous to ± appressed hirsute.
    Upper lvs. glabrous; phyllaries glabrous.
      E. inornatus var. *inornatus*
    Upper lvs. hirsute; phyllaries glandular.
      E. inornatus var. *viscidulous*
Heads radiate.
  Pappus distinctly double, with 5–12 inner bristles and a few outer paleae . . . . *E. divergens*
  Pappus not distinctly double, the bristles more numerous.
    Lf. blades glabrous.
      Heads mostly solitary; invol. ca. 8 mm high; pappus tawny . . . . *E. peregrinus* ssp. *callianthemus*
      Heads 1–8; invol. 4–6 mm high; pappus white.
        E. cervinus

Lf. blades ± hirsute. . . . . *E. eatonii* ssp. *plantagineus*

E. cervinus Greene. Siskiyou Daisy. Perennial from a branching, fibrous-rooted caudex; stems branched above, 1–2 dm tall, glabrous to loosely hirsute below, becoming more densely so near the summit of the peduncles; basal lvs. entire to slightly serrate, oblanceolate to elliptic, 1–4 cm long on broad ± ciliate petioles, cauline lvs. scattered and reduced, sessile; heads 4–12; invol. 4–6 mm high, the phyllaries narrowly oblanceolate, ± finely glandular, mostly green, sometimes purple at the tip; rays 20–50, blue to purple, disk corollas yellow; achenes ca. 2 mm long, light brown, 4-nerved, pubescent; pappus white, of many slender, minutely barbed bristles. Occasional. Open rocky places, 5,000–6,700 ft. Mixed Conifer F., Red Fir F. Boulder Cr. Basin (*176*), Caribou Basin (*I. L. Wiggins 13,518*). July–August.

E. divergens T. & G. Diffuse Daisy. Perennial from a slender taproot; stems decumbent to erect, 1–2.5 dm tall, densely hirsute throughout; basal lvs. oblanceolate to spatulate, petiolate, the blades 1–3 cm long, soon deciduous, cauline lvs. numerous, linear to linear-oblong, 1–2 cm long, densely hirsute, sessile; heads several; invol. ca. 5 mm high, phyllaries linear-ovate, hirsute and finely glandular, reflexed in age; rays blue, 75–150, disk corollas yellow; achenes ca. 1 mm long, light brown, sparsely pubescent; pappus double, of 5–12 inner minutely barbed bristles and a few outer short paleae. Found at one location. Dry, open woods, 5,000 ft. Mixed Conifer F. Big Flat (*385*). August.

E. eatonii Gray ssp. *plantagineus* (Greene) Cronq. Eaton's Daisy. Fig. 21. Perennial from a slender taproot; stems ± decumbent, 0.5–1.5 dm long, thinly hirsute with spreading to appressed hairs; basal lvs. linear to linear oblanceolate, up to 10 cm long, thinly hirsute, the cauline reduced; heads numerous; invol. 4–6 mm high, phyllaries narrowly oblanceolate, glandular-hirsute; rays 10–25, lavender to bluish, disk corollas yellow; achenes 1.5–2 mm long, light brown, sparingly pubescent; pappus tawny, of 15–30 slender, minutely barbed bristles. Occasional. Open, rocky meadows, 6,000–7,100 ft. Red Fir F. Ridge west of Bullards Basin (*968*), head of Union Cr. (*992*), Landers L. (*1469*). July–August.

E. inornatus (Gray) Gray var. *inornatus*. California Rayless Daisy. Stems 2–3 dm high; lvs. linear-oblanceolate, 3–5 cm long, reduced and glabrous above; phyllaries glabrous. Otherwise similar to the following var. Found at one location. Woods, 6,000 ft. Mixed Conifer F. North flank of Caribou Mt. (*1386*). August.

var. *visidulus* Gray. Perennial from a branched, woody caudex; stems numerous, slender, erect, 1–2 dm tall, densely hirsute with appressed hairs throughout; lvs. numerous, linear-oblong to linear-oblanceolate, 1–3 cm long, narrowed gradually to a sessile base, densely hirsute; heads 1–4, discoid; invol. 5–8 mm high, phyllaries lanceolate, glandular; corollas yellow; achenes 2–2.5 mm long, pale brown, sparingly pubescent, ± flattened; pappus sordid, of many barbed bristles. Occasional. Dry meadows and slopes, 6,500–7,300 ft. Red Fir F. Ward L. (*450*), Yellow Rose Mine (*954*) August.

E. peregrinus (Pursh) Greene ssp. *callianthemus* (Greene) Cronq. var. *angustifolius* (Gray) Cronq. Wandering Daisy. Fig. 22. Perennial from a short, stout, fibrous-rooted rhizome; stems erect, 1–3.5 dm tall, mostly simple, sparingly pubescent below, densely glandular-pubescent beneath the heads; lvs. oblanceolate, the basal 2–6 cm long on narrow petioles, the cauline shorter and sessile, glabrous; heads mostly solitary; invol. ca. 8 mm high, the phyllaries lanceolate and densely glandular; rays 15–30, blue to purple, disk corollas yellow; achenes densely pubescent; pappus tawny, of capillary, barbed bristles. Common. Damp slopes and meadows, 6,500–8,000 ft. Red Fir F., Subalpine F. Caribou Mt. (*210, 336*), Caribou Basin (*400,1046*), upper Canyon Cr. (*1315*). July–August.

E. petrophilus Greene. Rock Daisy. Perennial from a branched, woody rootcrown; stems numerous, erect, 1–3 dm tall, densely spreading-villous throughout; lvs. linear-oblanceolate, 1–2 cm long, numerous and fairly uniform, densely villous; heads 1–4, discoid, the corollas yellow; invol. 5–7 mm high, phyllaries lanceolate, densely glandular; achenes 2–2.5 mm long, pale brown, thinly pubescent; pappus sordid to ± tawny, of 20–40 barbed, capillary bristles. Occasional. Dry, open, rocky slopes, 5,200–7,500 ft. Montane Chaparral, Mixed Conifer F., Red Fir F. Canyon Cr. (*843*), Packers Pk. (*940*). August.

Eriophyllum Lag.

E. lanatum (Pursh) Forbes. var. *lanceolatum* (Howell) Jeps. Common Woolly-sunflower. Perennial from a branched, woody caudex, with a ± persistent white-woolly tomentum throughout; stems many, decumbent and spreading to erect, simple to few branched, 1–3 dm tall; lvs. 1–3 cm long, oblanceolate to obovate in outline, variously toothed or lobed; heads solitary or loosely corymbose on long, naked peduncles; invol. campanulate to hemispheric, 8–11 mm high; ray fls. 8–13, bright yellow, disk corollas yellow; achenes dark, 4-angled, slightly pubescent toward the summit, ca. 3.5 mm long; pappus of 4–6 white, erose, paleaceous scales ca. 1 mm long. Common. Dry, open, rocky places, 5,600–6,500 ft. Mixed Conifer F., Red Fir F. Boulder Cr. Basin (*177*), Canyon Cr. (*770*), Yellow Rose Mine (*957*), Head of Union Cr. (*991*). July–August.

Gnaphalium L.

G. palustre Nutt. Lowland Cudweed. Low annual from a delicate fibrous root system; stems erect, 5–10 cm high, densely white-woolly, especially above; lvs. spatulate to oblanceolate, 1–3 cm long, ± tomentose; heads several in tight clusters; invol. 2.5–3.5 mm high, densely woolly below, the phyllaries brownish with hyaline tips, and a green midrib below; corolla pale; achenes ca. 0.5 mm long, light brown; pappus white, of fine capillary bristles. Found at one location. Dry lake bed, 5,000 ft. Mixed Conifer F. Big Flat (*375*). August.

Haplopappus Cass.

Heads radiate; lvs. entire.
 Shrubs to 40 cm high . . . . *H. bloomeri*
 Low herbs to 15 cm high . . . . *H. lyallii*
Heads discoid; lvs. sharply toothed . . . . *H. whitneyi* ssp. *discoideus*

H. bloomeri Gray. Bloomer's Macronema. Low, broad shrub to 40 cm tall; stems resinous; lvs. numerous, nearly filiform to narrowly oblanceolate, 2–7 cm long, 1–2 mm wide, resinous; heads numerous; invol. narrowly companulate, ca. 1 cm high, phyllaries 3–6 seriate, the outer with caudate, herbaceous tips, at least the inner scarious margined, ciliate; rays 1–5, yellow, disk corollas 4–13, yellow; achenes 5.5–6.5 mm long, brown, angled, glabrate to sparingly pubescent below, densely so toward the summit; pappus sordid, of many unequal, minutely barbed, capillary bristles. Occasional. Open, gravelly places, 5,000 ft. Mixed Conifer F. Big Flat (*486*). August.

H. lyallii Gray. Lyall's Tonestus. Fig. 23. Cespitose perennial; stems 5–15 cm high, glandular-pubescent; lvs. oblanceolate, mucronate, glandular-pubescent, prominently veined, the lower petioled, 2–6 cm long, the upper sessile and reduced; heads solitary, 2 cm across; involucral bracts lanceolate, glandular-pubescent; ray and disk corollas yellow; pappus of minutely barbed bristles; achenes 4 mm long, sparingly pubescent in upper part. Rare. Dry, rocky ridges, 8,500–8,800 ft. Subalpine F. Thompson Pk. (*1308*), crest between Canyon Cr. and Rattlesnake Cr. (*J. O. Sawyer 2337*). July–August. This species was previously not reported for Calif. (Ferlatte, 1972).

H. whitneyi Gray ssp. *discoideus* (J. T. Howell) Keck. Whitney's Haplopappus. Woody subshrubs from a branching caudex; stems procumbent, 10–20 cm high, glandular-pubescent above; lvs. crowded, sessile, obovate, sharply toothed, 12–25 mm long, glandular-pubescent; heads crowded at the ends of the branches; invol. ca. 10 mm high; phyllaries glandular, with scarious margins; corollas yellow; pappus rusty-brown; achenes 5 mm long, prominently nerved, with a few scattered hairs. Found at one location. Dry, rocky ridge, 7,500 ft. Subalpine F. Ridge between Caribou Basin and Stuart Fork (*1382*). August–September.

Erigeron eatonii ssp. plantagineus

Erigeron peregrinus ssp. callianthemus

Haplopappus lyallii

Helenium bigelovii

Helenium L.

H. bigelovii Gray. Bigelow's Sneezeweed. Fig. 24. Perennial from a woody caudex; stems 3–7 dm tall, single or in clumps, simple to few branched, glabrous to puberulent; lvs. glandular-punctate, the lower petiolate, 3–12 cm long, oblanceolate to elliptic, the upper becoming reduced and sessile; heads mostly solitary on long, naked peduncles; phyllaries soon reflexed; rays 13–20, yellow, disk ± globose, corollas yellow; achenes ca. 1.5 mm long, brown, ribbed, ± pubescent; pappus of 6–8 lanceolate paleae, attenuate into an awn ca. half as long as disk corolla. Common. Bogs and wet meadows, 5,000–7,000 ft. Mixed Conifer F., Red Fir F. Big Flat (*231*), Dorleska Mine (*314*), Yellow Rose Mine (*947*), head of Union Cr. (*999*). August.

Hieracium L.

Corollas white to cream . . . . *H. albiflorum*
Corollas yellow.
  Lvs. pubescent with shaggy hairs . . . . *H. cynoglossoides*
  Lvs. glabrous . . . . *H. gracile*

H. albiflorum Hook. White-flowered Hawkweed. Perennial from a stout caudex bearing long fibrous roots; stems 1–several, 4–6 dm tall, hirsute below, glabrous above; basal lvs. linear-oblanceolate, 4–8 cm long, petiolate, covered with long white, shaggy hairs, the cauline lvs. becoming reduced, much less hairy and sessile; heads few to many; invol. 6–10 mm high, the phyllaries lanceolate, sparingly hirsute and glandular with narrow, scarious margins; rays white to cream; achenes brown, ca. 3 mm long; pappus tawny, of many minutely barbed bristles. Common. Open woods, 5,000 ft. Mixed Conifer F. Canyon Cr. (*457*), Big Flat (*705*). July–August.

H. cynoglossoides Arv.-Touv. ex Gray. Houndstongue Hawkweed. Perennial from a woody rootstock; stems slender, branched above, stellate-pubescent and glandular toward the heads; 1–3 dm high; lvs. numerous, basal, oblanceolate, 3–10 cm long, stellate-pubescent and with longer, shaggy hairs; invol. narrow, 7–11 mm high, phyllaries narrowly lanceolate, stellate-pubescent and with longer, black, glandular hairs; rays bright yellow; achenes reddish brown or darker, strongly ribbed, 3–3.5 mm long; pappus of barbed, tawny bristles. Common. Dry, open slopes, 5,000–8,000 ft. Mixed Conifer F., Red Fir F., Subalpine F. Big Flat (*252*), Josephine L. (*275*), Caribou Mt. (*340, 1041*), Canyon Cr. (*771*), north side of Thompson Pk. (*913*), Packers Pk. (*1398*). July–August.

H. gracile Hook. Alpine Hawkweed. Fig. 25. Stems 15–30 cm high, from a simple rootstock, becoming stellate and glandular toward heads; lvs. basal, narrowly oblanceolate to spatulate, 3–8 cm long, glabrous; heads several; involucre 5–10 mm high, densely hirsute with black, glandular hairs; corolla yellow; achenes reddish-brown, ribbed; pappus tawny. Common. Wooded slopes and damp, rocky places, 6,000–8,300 ft. Red Fir F., Subalpine F. Thompson Pk. (*1299*), Caribou Basin (*1380*). July–September.

Madia Mol.

Plants 1–5 dm tall; disk fls. numerous . . . . *M. gracilis*
Plants 0.5–1.5 dm tall; disk fls. 1 . . . . *M. minima*

M. gracilis (Sm.) Keck. Slender Tarweed. Tall, erect perennial; stems 1–5 dm tall, mostly simple below, hirsute to pubescent, becoming glandular above; lvs. linear, sessile, 1–4 cm long, up to 5 mm wide, densely covered with short, coarse hairs; heads several; invol. ovoid to depressed-globose, 6–9 mm high; the phyllaries covered with stout gland-tipped hairs; corollas yellow, the disk fls. numerous; ray achenes enclosed by phyllaries, 2.8–5 mm long, ± curved, often mottled, the disk achenes similar but straighter. Occasional. Dry, open slopes, 5,000 ft. Mixed Conifer F. Big Flat (*248*). August.

M. minima (Gray) Keck. Hemizonella. Low, ± spreading annual; stems divaricately branched, 4–15 cm tall, villous below, glandular-pubescent above; lvs. linear-oblong, 1–2 cm long, hirsute, often in clusters at the nodes; heads numerous, solitary or in small glomerules; invol. 2–3 mm high, the phyllaries loosely appressed, rounded on the back and covered with stout gland-tipped processes; corollas yellow, the disk fl. 1; ray achenes enclosed by phyllaries, incurved, ± mottled. Occasional. Dry, open slopes, 5,000 ft. Mixed Conifer F. Big Flat (*541*). August.

Microseris D. Don

M. nutans (Hook.) Sch.-Bip. Nodding Scorzonella. Perennial with milky juice from one to few fusiform taproots; stems 1–3 dm tall, glabrous; lvs. filiform to lance-linear, 8–15 cm long, glabrous, entire or toothed; heads solitary on the ends of branches, nodding in bud; invol. 10–15 mm high; outer phyllaries ciliolate and much shorter than the inner which may be pubescent on the margins and tips with black hairs; rays orangish yellow; achenes dark, 5–7 mm long; pappus white, of 15–20 narrow scales each bearing a capillary, plumose bristle. Occasional. Open, rocky meadows, 5,600–6,500 ft. Red Fir F. Boulder Cr. Basin (*172*), Kidd Cr. (*694*). July.

Raillardella Gray in Benth. & Hook.

R. pringlei Greene. Showy Raillardella. Fig. 26. Perennial from a slender rhizome; stems simple, 3–4 dm tall, becoming densely glandular above; lvs. glabrous, linear-oblanceolate, 3–10 cm long, basal and on lower stem, tending to be opposite; heads large, showy, solitary; invol. 12–15 mm high, phyllaries linear-lanceolate, ± ciliate, stipitate-glandular, purple toward the tips; corollas orange, the rays broad and deeply trifid; achenes pale, pubescent, ca. 7 mm long; pappus of ca. 15 flattened, plumose bristles. Rare. Bogs and wet places, 6,500–7,200 ft. Subalpine F. Head of Union Cr. (*998*), Landers L.

Basin (*1472*). July–August. This species has been previously reported only from Mt. Eddy (Munz, 1959).

Senecio L.

Cauline lvs. rapidly reduced upward; stems ± arachnoid-woolly. *S. integerrimus* var. *major*

Cauline lvs. only gradually reduced upwards; stems mostly glabrous. *S. triangularis*

S. integerrimus Nutt. var. *major* (Gray) Cronq. Single-stemmed Butterweed. Fig. 27. Stout perennial from a short crown with fibrous roots; stems solitary, 3–5 dm tall, thinly cobwebby, later ± glabrate; lower lvs. ovate to lance-oblong, denticulate, the blades 4–10 cm long on petioles from much shorter to ca. as long, cauline lvs. becoming sessile and reduced upwards; heads many, in a ± flat-topped, open cluster; invol. 6–10 mm high; phyllaries acuminate, ± cobwebby; corollas yellow; achenes brown, ca. 4 mm long; pappus white, of many fragile, capillary bristles. Common. Woods and slopes, 5,000–6,000 ft. Mixed Conifer F., Red Fir F. Yellow Rose Mine (*129*), Big Flat (*147*, *539*), Caribou Basin (*197*). June–July.

S. triangularis Hook. Arrowhead Butterweed. Stout perennial from a fibrous-rooted crown; stems solitary to few, 3–7 dm tall, glabrous; lvs. numerous, sharply toothed, gradually reduced upwards, the lower triangular-ovate to lanceolate, the blades 4–8 cm long on long petioles, the upper narrower on shorter petioles or sessile; heads numerous in a ± flat-topped cluster; invol. 4–6 mm high; phyllaries slightly acuminate, glabrous except for tuft of hairs at tips; corollas yellow; achenes brown, ca. 4 mm long; pappus white, of many fragile capillary bristles. Common. Wet meadows and stream banks, 6,000–7,000 ft. Red Fir F., Subalpine F. Caribou Mt. (*323*), Canyon Cr. (*798*), Landers L. (*1011*), Kidd Cr. (*1111*). August.

Solidago L.

S. canadensis L. ssp. *elongata* (Nutt.) Keck. Meadow Goldenrod. Fig. 28. Perennial from creeping rhizomes; stems 1.5–5 dm tall, puberulent to below the middle, often reddish below; lvs. numerous, lanceolate to lance-oblong, 1–5 cm long, entire to sharply serrate; heads small and numerous in an elongate, open or short, congested infl.; invol. ca. 3 mm high, 2–3 seriate; corollas yellow; achenes hispidulous, 0.5–1 mm long; pappus white, of many fragile, capillary bristles. Common. Meadows, stream banks, and open slopes, 5,200–7,000 ft. Mixed Conifer F., Red Fir F. Josephine L. (*290*), Dorleska Mine (*310*), South Fork of Salmon R. (*415*), Grizzly L. Basin (*880*), north side of Red Rock Mt. (*978*). August.

Stephanomeria Nutt.

S. lactucina Gray. Large-flowered Stephanomeria. Perennial with milky juice; from a slender rootstock; stems simple or paniculately branched, 6–15

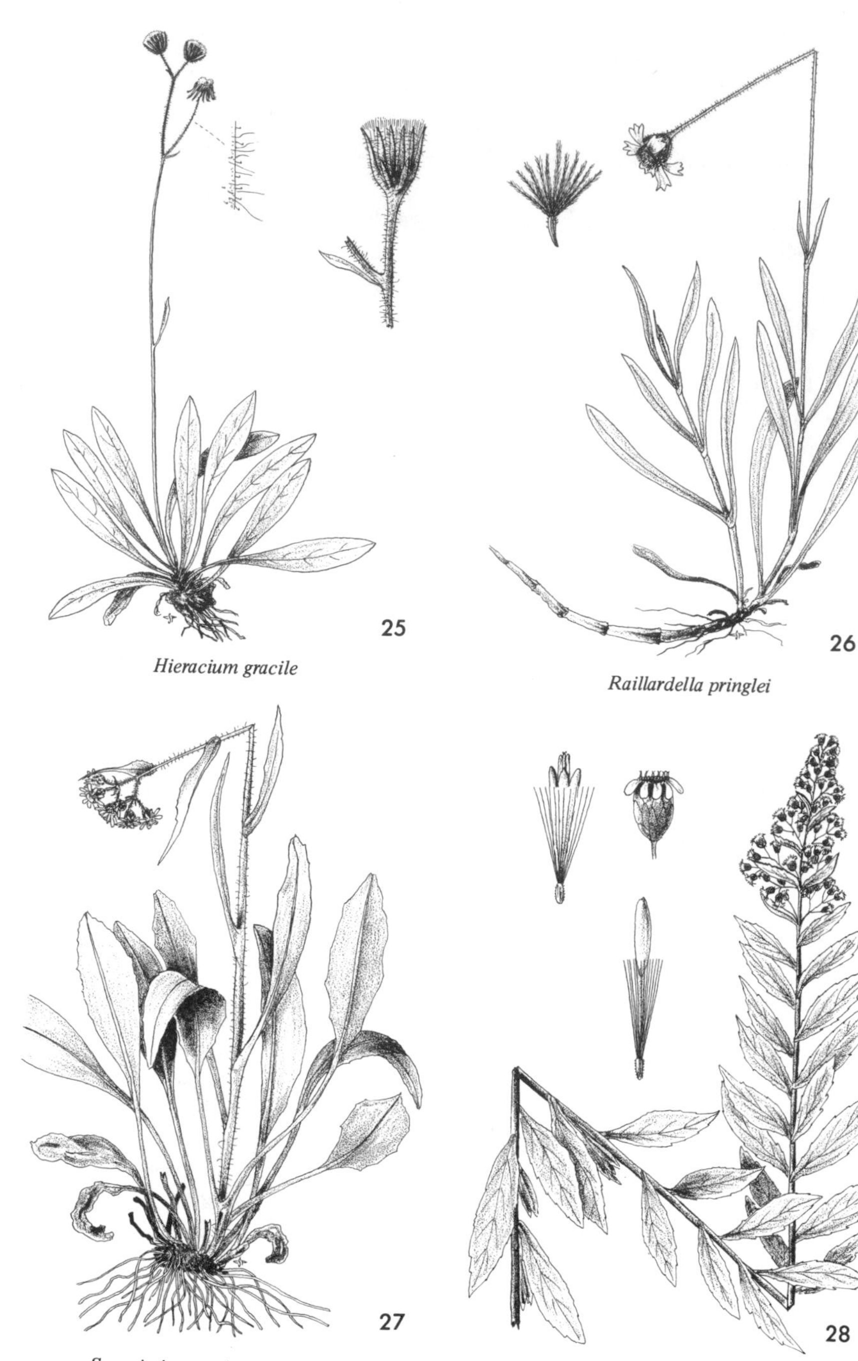

Hieracium gracile

Raillardella pringlei

Senecio integerrimus var. major

Solidago canadensis ssp. elongata

cm tall, puberulent; lvs. linear, 1–7 cm long, up to 4 mm wide, entire or with a few teeth, glabrous; heads few, terminal on the ± corymbosely arranged peduncles; invol. 12–15 mm high, the phyllaries ± puberulent; rays lavender or pink; achenes 5–6 mm long, ± angled; pappus white, of ca. 20 finely plumose bristles ± united at base. Occasional. Woods and dry slopes, 5,000–7,200 ft. Mixed Conifer F., Red Fir F. Caribou Mt. (*327, 1081*), Big Flat (*363*). August.

Taraxacum Wiggers

T. officinale Wiggers. Common Dandelion. Scapose perennial from a stout taproot; scapes 5–60 cm high, glabrous below, sparingly pubescent above; lvs. oblanceolate 5–35 cm long, toothed to deeply lobed; phyllaries strongly reflexed at maturity, ca. 15 mm long; corolla yellow; achenes greenish brown, 4 mm long, ribbed and bearing sharp tubercles toward summit; pappus white, borne on a slender beak 6–9 mm long. Occasional. Around meadows, 5,000–7,000 ft. Mixed Conifer F., Red Fir F. Head of Kidd Cr. (*1403*). June–July.

Wyethia Nutt.

W. angustifolia (DC.) Nutt. Narrow-leaved Mule-ears. Stout perennial from a thick taproot; stems 3–5 dm tall, densely hirsute; basal lvs. ca. equaling the stems, tapering at base and apex, long-petiolate, resinous and hispid, cauline lvs. reduced and becoming sessile; heads 1–4, large and showy; invol. 1.5–2.5 cm high, the phyllaries hirsute and strongly ciliate; corollas yellow; receptacle bearing large, pubescent paleae which nearly surround the achenes; achenes dark, 7–9 mm long; pappus of short, united scales extended into one or more stout, scabrous awns. Found at one location. Open meadow, 5,000 ft. Mixed Conifer F. Big Flat (*262*). August.

CONVOLVULACEAE

Convolvulus L.

Lvs. tomentose . . . . *C. malacophyllus*
Lvs. glabrous . . . . *C. occidentalis* var. *solanensis*

C. malacophyllus Greene. Sierra Morning-glory. Fig. 29. Low, trailing, tomentose perennial; stems 1–3 dm long; lvs. triangular-hastate, 1–2.5 cm long, up to 4 cm broad, the basal lobes 1 or 2 toothed, petioles 4–20 mm long; fls. borne singly on axillary peduncles 1–3 cm long; 2 ovate bracts 10–12 mm long closely subtend the fls. and ± conceal the calyx; sepals lance-ovate, mucronate, ca. 10 mm long; corolla cream to pale yellow, 2–2.5 cm long. Occasional. Dry slopes, 5,000–5,500 ft. Montane Chaparral. Yellow Rose Mine Trail (*616*). July.

C. occidentalis Gray var. *solanensis* (Jeps.) J. T. Howell. Western Morning-glory. Perennial, trailing over trees and shrubs; lvs. glabrous, 1–4 cm long, hastate, the basal lobes toothed; peduncles 1–3 fld.; bracts 5–15 mm below fls.; sepals 6–10 mm long; corolla cream to yellowish, ca. 3 cm long. Found at one location. Brushy slope, 5,300 ft. Mixed Conifer F. Boulder Cr. Basin (*1279*). July–August.

CORNACEAE

Cornus L.

Fls. in dense heads subtended by white, petallike bracts. . . . *C. nuttallii*
Fls. in bractless cymes. . . . . *C. stolonifera*

C. nuttallii Aud. Mountain Dogwood. Spreading tree 3–4 m tall; lvs. opposite, elliptic to obovate, 6–10 cm long, strigulose above, paler beneath and pubescent with appressed hairs, petioles 5–8 mm long, strigulose; fls. in dense heads subtended by 4–7 white, petaloid bracts, 4–5 cm long; calyx 2.5 mm high; petals whitish, 3–4 mm long; style ca. 2 mm long; drupes red, 1–1.5 cm long. Found at one location. Open woods, 5,400 ft. Mixed Conifer F. Canyon Cr. (*840*). July–August.

C. stolonifera Michx. American Dogwood. Fig. 30. Spreading shrub to 2 m tall; lvs. opposite, elliptic, 2–10 cm long, pubescent, paler beneath, petioles 5–25 mm long, strigulose; fls. many in a compact cyme; sepals 0.5 mm long; petals white ca. 3 mm long; style ca. 2 mm long; drupe white, 7–9 mm in diam. Found at one location. Stream bank, 5,400 ft. Mixed Conifer F. Canyon Cr. (*839*). August.

CRASSULACEAE

Sedum L.

Lvs. of flowering stem much reduced and different than basal lvs.
S. obtusatum ssp. *boreale*
Lvs. of flowering stems not much different than basal lvs.
S. stenopetalum ssp. *radiatum*

S. obtusatum Gray ssp. *boreale* Clausen. Sierra Stonecrop. Fig. 31. Perennial from creeping rootstocks; flowering stem solitary, 10–15 cm tall; lvs. succulent, the basal 1–2.5 cm long, green to reddish, entire to retuse, the cauline reduced, 0.5–1 cm long; infl. a dense, many fld. panicle; sepals ca. 2 mm long; petals yellow, united at the base, erect to spreading, 5–7 mm long; follicles erect to slightly divergent, 5–6 mm long; style slender, ca. 2 mm long. Common. Dry, rocky slopes, 5,500–8,000 ft. Red Fir F., Subalpine F. Josephine L. Basin (*113*), Canyon Cr. (*167*), Caribou Mt. (*332*), head of Kidd Cr. (*455*), Dorleska Summit (*612*), Caribou Basin (*750*), Grizzly L. Basin (*898*), Red Rock Mt. (*1029*). June–August.

S. stenopetalum Pursh. ssp. *radiatum* (Wats.) Clausen. Narrow-petaled Stonecrop. Perennial from creeping rootstocks; flowering stems ca. 10 cm tall; lvs. succulent, the basal crowded, 3–4 mm long, the cauline scattered, similar to basal; infl. several to many fld.; sepals 3–4 mm long; petals yellow, united at base, somewhat spreading, ca. 7 mm long; follicles erect to spreading, 4–5 mm long; styles slender, ca. 1 mm long. Found at one location. Open, rocky slope, 7,000 ft. Subalpine F. Grizzly L. Basin (895). August.

CRUCIFERAE

Fruit a silicle, only 1–3 times longer than wide.
 Silicles 1–2.5 mm long, covered with hooked hairs...... *Athysanus*
 Silicles mostly longer, not covered with hooked hairs.
  Silicles compressed parallel to the broad partition.
   Silicles densely stellate-pubescent; fruiting pedicels often sigmoid .... *Lesquerella*
   Silicles glabrous to thinly stellate-pubescent; fruiting pedicels usually straight.......................... *Draba*
  Silicles compressed contrary to the narrow partition.
   Silicles triangular .... *Capsella*
   Silicles not triangular.
    Stems and lvs. densely short-hispid...... *Lepidium*
    Stems and lvs. glabrous.................. *Thlaspi*
Fruit a silique, 4 or more times longer than wide.
 Petals yellow or orange.
  Lower lvs. entire to denticulate .... *Erysimum*
  Lower lvs. ± pinnately divided.
   Siliques 25–35 mm long.................... *Barbarea*
   Siliques 9–15 mm long.
    Stems and lvs. densely stellate-pubescent. *Descurainia*
    Stems and lvs. glabrous to sparingly hispid.. *Rorippa*
 Petals white to deep purple.
  Plants glabrous.
   Petals purple .... *Arabis*
   Petals white.
    Cauline lvs. strongly auriculate-clasping. *Streptanthus*
    Cauline lvs. not auriculate-clasping ...... *Dentaria*
  Plants pubescent, at least on the lvs.
   Caudex thick, heavy, and covered with remains of dead lvs. .... *Phoenicaulis*
   Caudex not as above........................ *Arabis*

Arabis L.

Fruiting pedicels erect to ascending.
 Stems and upper lvs. glabrous.
  Basal lvs. glabrous except for base of petioles ...... *A. lyallii*
  Basal lvs. sparsely pubescent with dendritic hairs throughout. *A. platysperma* var. *howellii*

Stems and upper lvs. pubescent with dendritic hairs.
Siliques 1–2 mm wide . . . . *A. breweri*
Siliques 3–4 mm or more wide. . . . . *A. platysperma*
Fruiting pedicels widely spreading to strongly reflexed.
Cauline lvs. glabrous.
Caudex branched; siliques 2–3 mm wide . . . . *A. suffrutescens*
Caudex simple; siliques 1 mm wide. . . . . *A. rectissima*
Cauline lvs. densely stellate-pubescent.
Fruiting pedicels mostly widely spreading. . . . . . *A. sparsiflora*
Fruiting pedicels arched downward to strongly reflexed.
Pedicels geniculate . . . . *A. holboellii* var. *retrofracta*
Pedicels not geniculate.
Siliques blunt at apex; styles obsolete. . . *A. puberula*
Siliques acuminate; style 1–2 mm long.
A. subpinnatifida

A. breweri Wats. Brewer's Rock-cress. Perennial from a branched, woody caudex; stems several, 3–20 cm tall, hirsute below becoming sparsely so above; basal lvs. linear-spatulate, 1–2 cm long, densely pubescent with dendritic hairs, entire to crenate, cauline lvs. sessile, auriculate, ca. 1 cm long, glabrate above, pubescent below; sepals purple to greenish, 2.5–3 mm long, pubescent with dendritic hairs; petals purple, 6–7 mm long; fruiting pedicels ascending, sparsely pubescent, 4–10 mm long; siliques arcuate to straight, 3–5 cm long, 1–2 mm wide, glabrous. Occasional. Dry, rocky slopes, 5,500–7,000 ft. Red Fir F. Yellow Rose Mine Trail (*134*), Packers Pk. (*549*), Grizzly L. Basin (*901*). June.

A. holboellii Hornem. var. *retrofracta* (Grah.) Rydb. Holboell's Rock-cress. Fig. 32. Perennial from a simple or branched caudex; stems 1 to few, 15–40 cm tall, densely stellate-pubescent below, becoming sparsely so to glabrous above; basal lvs. spatulate to linear-oblong, 1–2 cm long, densely stellate-pubescent, cauline lvs. sessile, 1–2 cm long, auriculate, densely stellate-pubescent; sepals green to purple, 2.5–3 mm long, densely hirsute with dendritic hairs; petals whitish, 4–6 mm long; fruiting pedicels strongly reflexed, appressed to stem, geniculate, mostly densely stellate-pubescent, 6–10 mm long; siliques straight, 3–5 cm long, 1–1.5 mm wide, ± appressed to stem, glabrous to stellate-pubescent. Common. Dry, rocky slopes and meadows, 5,000 ft. Mixed Conifer F. Canyon Cr. (*67*), Big Flat (*87, 517*). June.

A. lyallii Wats. Lyall's Rock-cress. Low perennial from a simple to branched caudex; stems 1 to several, 4–13 cm tall, glabrous; basal lf. blades narrowly oblanceolate, 1–2 cm long, glabrous, on slender petioles with a few dendritic hairs toward the base, cauline lvs. linear to lanceolate, ca. 1 cm long, sessile, ± auriculate-clasping, glabrous; sepals greenish with purple tips, and upper margins, ca. 3 mm long, glabrous; petals purple, 6–7 mm long; fruiting pedicels erect to ascending, 4–8 mm long, glabrous; siliques erect or nearly so, 3–5 cm long, 2–3 mm wide, glabrous. Occasional. Damp, rocky slopes, 7,400–7,800 ft. Subalpine F., Alpine Fell-fields. Below ice field on north side of Thompson Pk. (*923*), Caribou Basin (*1052*). August.

A. *platysperma* Gray. Broad-seeded Rock-cress. Perennial from a branching caudex; stems usually several, 10–20 cm tall, stellate-pubescent below becoming glabrate above; basal lvs. oblanceolate, 1–3 cm long, stellate-pubescent, cauline lvs. few, scattered, oblong to linear-lanceolate, ca. 1 cm long, sessile, not auriculate, pubescent with dendritic hairs; sepals greenish to purple, 3–4 mm long, glabrous; petals purple to whitish, 4–6 mm long; fruiting pedicels erect to ascending, 3–6 mm long, fairly stout, ± pubescent with dendritic hairs; siliques straight, 3–5 cm long, 3–4 mm wide, flattened, glabrous. Occasional. Dry, wooded ridges, 6,000–8,000 ft. Red Fir F., Subalpine F. Sawtooth Ridge (*1071*), Boulder Cr. Basin (*1423*). August.

var. *howellii* (Wats.) Jeps. Glabrous except for basal lvs. which are sparsely pubescent with dendritic hairs; cauline lvs. auriculate. Occasional. Dry, rocky ridges, 6,500–8,100 ft. Red Fir F., Subalpine F. Caribou Mt. (*185*, *1374*). July–August.

A. *puberula* Nutt. Blue Mountain Rock-cress. Perennial from a simple caudex; stems solitary, 20–30 cm tall, hoary with a dense stellate-pubescence below, becoming green and sparsely pubescent above; basal lvs. oblanceolate, 1–2 cm long, hoary with a dense, fine stellate-pubescence, cauline lvs. linear, 1–2 cm long, sessile, hoary; sepals purple, 4–6 mm long, densely hirsute with dendritic hairs; petals purple to whitish, 6–7 mm long; fruiting pedicels strongly reflexed, 4–7 mm long, hirsute with dendritic hairs; siliques pendulous, 3–6 cm long, 1.5–2 mm wide, stellate-pubescent, especially toward the base. Found at one location. Dry, rocky ridge, 7,000 ft. Subalpine F. Dorleska Summit (*608*). July.

A. *rectissima* Greene. Bristly-leaved Rock-cress. Perennial from a simple caudex; stems glabrous, 10–20 cm tall; basal lvs. spatulate to oblanceolate, 1–2 cm long, ciliate with forked hairs, otherwise glabrous, cauline lvs. linear to oblanceolate, 1–1.5 cm long, glabrous, ± auriculate-clasping; sepals green, 3–4 mm long, glabrous; petals whitish to purple, 6–7 mm long; fruiting pedicels reflexed, 4–12 mm long, glabrous; siliques mostly straight, reflexed, 2–3 cm long, ca. 1 mm wide. Found at one location. Dry, rocky ridge, 7,000 ft. Subalpine F. Dorleska Summit (*609*). July.

A. *sparsiflora* Nutt. in T. & G. Elegant Rock-rose. Perennial from a simple caudex; stems solitary, ca. 60 cm tall, simple below, with several slender branches above, hirsute with dendritic hairs below, becoming very sparsely so

Convolvulus malacophyllus

Cornus stolonifera

Sedum obtusatum ssp. boreale

Arabis holboellii var. retrofracta

above; basal lvs. linear-oblanceolate, 3–10 cm long, pubescent with dendritic hairs, cauline lvs. numerous, lanceolate, 2–3.5 cm long, densely stellate-pubescent, ± auriculate-clasping; sepals 4–6 mm long; petals pink to purple, 8–14 mm long; fruiting pedicels spreading to slightly reflexed or ascending, 4–5 mm long, sparsely hirsute with spreading, dendritic hairs; siliques mostly arcuate, 4–7 cm long, 1.5–2 mm wide, spreading to slightly ascending, mostly glabrous. Found at one location. Dry, rocky slope, 6,500 ft. Red Fir F. Northwest side of Caribou Mt. (*763*). July.

A. *subpinnatifida* Wats. Klamath Rock-cress. Perennial from a branched caudex; stems 1–several, 15–25 cm tall, densely stellate-pubescent below, becoming somewhat less so above; basal lvs. oblanceolate, slightly toothed to subpinnatifid, mostly ca. 1 cm long, densely pubescent with fine, stellate hairs, cauline lvs. lanceolate, mostly subpinnatifid, 1–1.5 cm long, densely pubescent with fine, stellate hairs; sepals mostly greenish, 5–6 mm long, stellate-pubescent; petals purple to lavender, 10–12 mm long; fruiting pedicels arched downward, 10–13 mm long, stellate-pubescent; siliques reflexed to pendulous, 5–7 mm long, 2–3 mm wide, acuminate at the tip, glabrous; style 1–2 mm long. Found at one location. Dry rocky ridge, 7,000 ft. Red Fir F. Packers Pk. (*558*). June–July.

A. *suffrutescens* Wats. Woody Rock-cress. Glaucous perennial from a branched, woody caudex; stems glabrous, 20–40 cm high; basal lvs. spatulate to oblanceolate, 1–2 cm long, ciliate with forked hairs, the surfaces glabrous to sparingly pubescent, cauline lvs. linear to lanceolate, 1–1.5 cm long, glabrous, auriculate; sepals 4 mm long; petals purple, 6–8 mm long; siliques reflexed, strongly flattened, acuminate, 4–6 cm long, 2–3 mm wide. Occasional. Rocky places, 5,000–8,000 ft. Mixed Conifer F., Red Fir F. Black Mt. (*1400*), Big Flat (*J. T. Howell 13,204*). July–August.

Athysanus Greene

A. *pusillus* (Hook.) Greene. Dwarf Athysanus. Diffuse, slender-stemmed annual, branching from the base; stems several, 10–20 cm tall, pubescent with forked hairs below, becoming glabrous above; lvs. mostly near base, ovate, 5–10 mm long, entire to few-toothed, densely stellate-pubescent; infl. an elongate, unilateral raceme; fls. inconspicuous, ca. 1.5 mm long; fruiting pedicels recurved; silicles 1–2.5 mm long, uncinate-hispid. Found at one location. Dry grassy woods, 5,000 ft. Mixed Conifer F. Big Flat (*499*). June.

Barbarea R. Br.

B. *orthoceras* Ledeb. American Winter Cress. Perennial from a simple to branched caudex; stems one to several, usually branched above, 1–3 dm tall; basal lvs. petioled, 3–7 cm long, divided into several pairs of small lfts. and one large terminal one, cauline lvs. lyrate-pinnatifid, 2–5 cm long; racemes dense at first, becoming more open in fruit; sepals yellow-green, 3–4 mm long; petals mostly yellow, sometimes purplish, 4–7 mm long; fruiting pedicels ascending, stout, 3–4 cm long, siliques 2.5–3.5 cm long, 1–1.5 mm wide. Occasional. Open meadows, 5,000 ft. Mixed Conifer F. Big Flat (*94, 591*). June–July.

Capsella Medic.

C. bursa-pastoris (L.) Medic. Shepherd's Purse. Slender stemmed annuals 1.5–2.5 dm high in fr.; rosette lvs. pinnatifid, pubescent with forked hairs, 3–10 cm long; cauline lvs. reduced and sessile with clasping bases; fls. ca. 2 mm long, petals white; silicles triangular-obovate, ca. 5 mm long, shallowly notched at summit, borne on slender, spreading pedicels 8–10 mm long, style ca. 0.2 mm long. Occasional. Disturbed areas, 5,000 ft. Mixed Conifer F. Big Flat (*1139*). June.

Dentaria L.

Rhizomal lvs. pinnately compound. *D. californica*
Rhizomal lvs. simple . . . . *D. californica* var. *cardiophylla*

D. californica Nutt. California Toothwort. Glabrous perennial from deep-seated ovoid rhizomes; stems mostly simple, 2–3.5 dm tall; rhizomal lvs. 3–5 foliate, the lfts. ovate, sinuate to dentate, 1–2 cm long, cauline lvs. mostly smaller; sepals green, ca. 3 mm long; petals white, 9–14 mm long; fruiting pedicels slender, erect, 1–2 cm long; siliques 2–5 cm long, 1–2 mm wide; style 2.5–4 mm long. Occasional. Bogs and meadows, 6,500 ft. Red Fir F. Dorleska Mine (*623*). July.

var. *cardiophylla* (Greene) Detl. Rhizomal and cauline lvs. simple, ovate, 2.5–4 cm long, coarsely toothed. Found at one location. Open, damp meadow, 6,100 ft. Red Fir F. Head of Union Cr. (*651*). July.

Descurainia Webb & Berthel.

D. richardsonii (Sweet) O. E. Schulz ssp. *viscosa* (Rydb.) Detl. Mountain Tansy-mustard. Annual or biennial from a simple, stout caudex; stems several, 1–3 dm tall, simple or few branched above, densely stellate-pubescent below, becoming less so and mixed with glandular hairs above, mostly purplish below; lvs. 1–4 cm long, densely stellate-pubescent, pinnately divided, the divisions of the lower lvs. broad, those of the upper lvs. becoming finer; sepals 1.5–2 mm long, pubescent; petals yellow, ca. 2 mm long; fruiting pedicels ascending, 5–10 mm long, slender, stellate-pubescent and ± glandular; siliques 9–15 mm long, ca. 1 mm wide, glabrous, torulose; style 0.5 mm long. Found at one location. Dry slope, 6,600 ft. Red Fir F. Near Dorleska Mine (*309*). August.

Draba L.

Lvs. densely stellate-pubescent; siliques planar. . . . . *D. howellii*

Lvs. thinly stellate-pubescent; siliques often curved or twisted.
D. howellii var. *carnosula*

D. howellii Wats. Howell's Draba. Fig. 33. Cespitose perennial from a branched, ± decumbent caudex; fruiting stems 4–10 cm tall, softly pubescent with simple and forked hairs throughout; lvs. obovate, 4–6 mm long, 2–3 mm wide, densely stellate-pubescent, 1–3 small bracts present on the flowering stem; infl. ca. 15 fld.; sepals ca. 2 mm long, pubescent with forked hairs; petals yellow, 2–4 mm long; fruiting pedicels ascending, 5–10 mm long, slender, pubescent with forked hairs; silicle ovate-elliptic, 5–8 mm long, pubescent with forked hairs; styles ca. 1.2 mm long. Occasional. Open, rocky slopes, 7,800–8,700 ft. Alpine Fell-fields. Below ice field on the north side of Thompson Pk. (*914*), near summit of Thompson Pk. (*1300*). August.

var. *carnosula* (O. E. Schulz) C. L. Hitchc. Cespitose perennial from a loosely branched caudex; fruiting stems 5–8 cm tall, softly pubescent with forked and simple hairs throughout; lvs. obovate to spatulate, 5–25 mm long, 2–12 mm wide, ciliate with stalked, stellate hairs and thinly stellate-pubescent on both surfaces, 1–3 small bracts present on the flowering stems; infl. 10–25 fld.; sepals 2–3 mm long, with a few simple or forked hairs; petals yellow, 4–6 mm long; fruiting pedicels widely spreading to ± ascending, 6–11 mm long, slender, thinly pubescent with simple and forked hairs; silicles ovate-elliptic, 3–7 mm long, thinly pubescent with forked hairs, often curved or twisted; styles ca. 1.5 mm long. Found at one location. Wet, rocky stream bank, 7,000 ft. Subalpine F. South end of Caribou Basin (*757, 1055*). July–August.

Erysimum L.

Siliques rounded to 4-angled; foliar hairs 2-, 3-, and 4-parted.
E. capitatum
Siliques flattened; foliar hairs mostly 2-parted .... *E. perenne*

E. capitatum (Dougl.) Greene. Douglas' Wallflower. Fig. 34. Perennial from a simple caudex; stems 1 to few, 15–40 or more cm tall, pubescent with mostly 2-parted hairs; basal lvs. many, crowded, ± rosulate, petiolate, the blades narrowly elliptic to oblanceolate, 1–6 cm long, 3–15 mm wide, denticulate, pubescent with 2-, 3-, and 4-parted hairs, cauline lvs. smaller and ± scattered; sepals ca. 10 mm long, pubescent toward the tip; petals yellow to orangish, 15–20 mm long; fruiting pedicels ascending, ca. 10 mm long, pubescent; siliques rounded to ± 4-angled, 3–6 cm long, 1.5–2 mm wide, pubescent. Common. Dry, open slopes, 6,500–8,100 ft. Red Fir F., Subalpine F. Dorleska Summit (*119*), Caribou Summit (*184*), north side of Caribou Mt. (*214*), Canyon Cr. (*811*), Grizzly L. Basin (*891*). June–August.

E. perenne (Wats. ex Cov.) Abrams. Sierra Wallflower. Perennial from a branched caudex; stems few, 15–20 cm tall, pubescent with 2–3 parted hairs;

basal lvs. linear-oblanceolate on slender petioles, the blades 1–2 cm long, 2–6 mm wide, ± denticulate, pubescent with mostly 2-parted hairs; sepals 7–9 mm long, pubescent; petals yellow, 15–17 mm long; fruiting pedicels ascending, 5–7 mm long, pubescent; siliques flattened, 3–6 mm long, 1.5–2 mm wide, pubescent. Found at one location. Dry, wooded slope, 7,500 ft. Subalpine F. Sawtooth Ridge (*1070*). August.

Lepidium L.

L. campestre (L.) R. Br. English Pepper-grass. Densely short-hispid annual or biennial; stems simple to few branched, 10–15 cm tall; lvs. oblong-oblanceolate, 1–2 cm long, ± denticulate, the cauline lvs. erect, ± overlapping and with auriculate-clasping bases; sepals 1–1.5 mm long, pubescent; petals white, ca. 2 mm long; fruiting pedicels ascending, 3.5–5 mm long; silicles 4–5 mm long, winged along the margins and especially at the apex, papillose, the cells each with one seed; styles ca. 0.3 mm long. Found at one location. Weedy area, 5,000 ft. Mixed Conifer F. Big Flat (*255*). July–August.

Lesquerella Wats.

L. occidentalis Wats. Western Lesquerella. Perennial from a stout, branched caudex, the entire plant densely silvery stellate-pubescent; stems several, 6–10 cm tall; basal lvs. spatulate, 2–5 cm long, cauline lvs. few, smaller; sepals 5–6 mm long; petals 6–8 mm long; fruiting pedicels ± sigmoid, 7–12 mm long; silicles 3–5 mm long, ± inflated; styles 4–5 mm long. Found at one location. Dry rocky ridge, 7,000 ft. Subalpine F. Dorleska Summit (*575*). June–July.

Phoenicaulis Nutt.

P. cheiranthoides Nutt. in T. & G. Common Phoenicaulis. Scapose perennial from a heavy caudex that is covered with old lf. bases; stems several, 10–15 cm tall, mostly glabrous; basal lvs. spatulate to oblanceolate, 3–5 cm long, stellate-tomentose, cauline lvs. smaller, auriculate-clasping; sepals yellowish green, 3–4 mm long, stellate-pubescent at the tips; petals purple, ca. 7 mm long; fruiting pedicels widely spreading to somewhat reflexed, 10–14 mm long; siliques strongly flattened, ± lanceolate, 2–3 cm long, 1.5–4 mm wide, glabrous. Found at one location. Dry rocky ridge, 7,800 ft. Subalpine F. Packers Pk. (*563*). June–July.

Rorippa Scop.

R. curvisiliqua (Hook.) Bessey. Western Yellow-cress. Annual or biennial from a branched caudex; stems many, diffusely branched and ± spreading, 6–12 cm long, sparingly hispid; lvs. 3–6 cm long, pinnatifid into usually acute, toothed lobes; sepals greenish yellow, 1–2 mm long; petals somewhat shorter, yellow; fruiting pedicels spreading to ascending, 1.5–3 mm long; siliques turgid, ± curved, 6–8 mm long, 1–1.5 mm wide; styles nearly obsolete. Found at one location. Dry lake bed, 5,000 ft. Mixed Conifer F. Big Flat (*365*). August.

Streptanthus Nutt.

S. *tortuosus* Kell. Mountain Streptanthus. Fig. 35. Glabrous perennial from a branched ± woody caudex; stems several, 15–25 cm tall; lower lvs. spatulate, 1–3 cm long, cauline lvs. oblanceolate to nearly round, strongly auriculate-clasping; sepals greenish yellow, 5–6 mm long; siliques arcuate-spreading, 5–8 cm long, 1–1.5 mm wide, glabrous. Common. Dry rocky slopes, 5,600–7,800 ft. Montane Chaparral, Red Fir F. Josephine L. Basin (*79*), Canyon Cr. (*772*), Black Mt. (*1084*). June–August.

Thlaspi L.

T. *glaucum* A. Nels. var. *hesperium* Pays. Penny Cress. Loosely cespitose perennial from a branched caudex; stems many, mostly simple, 4–15 cm tall; basal lvs. spatulate to oblanceolate, 5–12 mm long, cauline lvs. scattered, 5–7 mm long, auriculate-clasping; sepals 1.5–2 mm long; petals white to cream, 4–5 mm long; fruiting pedicels ± spreading to ascending, slender, ca. 5 mm long; silicles obovate, cuneate at base, 3–6 mm long; styles 1.5–2 mm long. Common. Dry slopes and meadows, 5,000–7,000 ft. Mixed Conifer F., Red Fir F. Yellow Rose Mine (*122*), Big Flat (*585*), Union L. (*644*), Landers L. (*985*). June-August.

ERICACEAE

Petals separate; capsule splitting from base upward . . . . *Ledum*
Petals connate; capsule splitting from apex downward, or fr. a berry.
Corollas white.
Lvs. scalelike, overlapping, sessile; corolla campanulate . . . *Cassiope*
Lvs. broad, petioled; corolla urceolate.
Ovary superior.
Fls. in elongate racemes, fr. a capsule . . . . *Leucothoe*
Fls. in compact racemes or looser panicles, fr. a berry . . . . *Arctostaphylos*
Ovary inferior . . . . *Vaccinium*
Corollas rose to purple.
Corolla saucer-shaped, 8–12 mm broad; lvs. broad . . . . *Kalmia*
Corolla campanulate, less than 8 mm broad; lvs. linear, needle-like . . . . *Phyllodoce*

Arctostaphylos Adans.

Stout shrubs, the branches spreading from an enlarged rootcrown; fruit 7–10 mm broad . . . . *A. patula*
Low, decumbent shrubs, the branches rooting in contact with the ground; fr. 5–6 mm broad . . . . *A. nevadensis*

Draba howellii

Erysimum capitatum

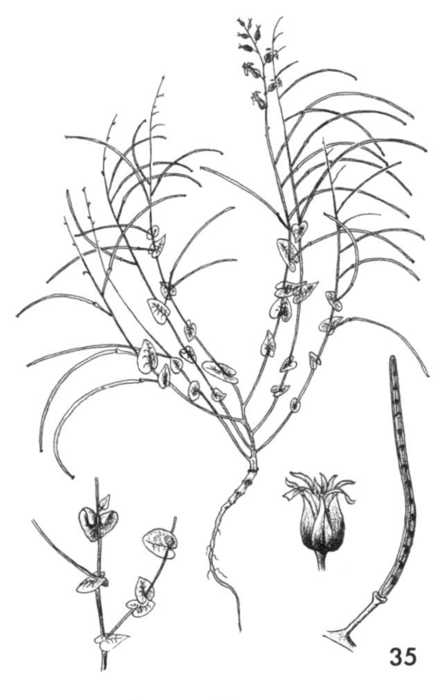

Streptanthus tortuosus

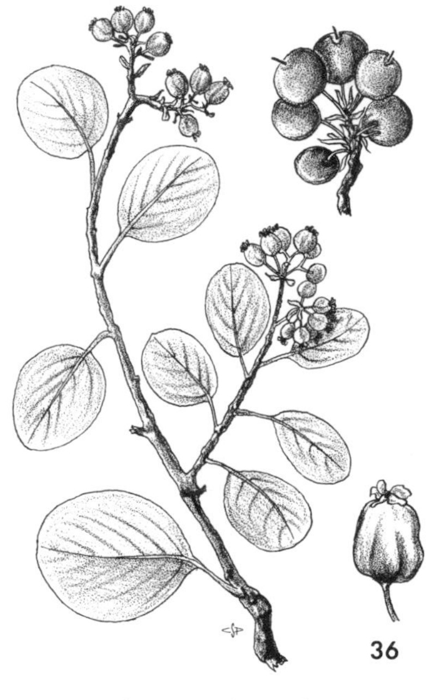

Arctostaphylos patula

A. nevadensis Gray. Pinemat Manzanita. Low, decumbent shrub, 1–3 dm high, the branches freely rooting; bark reddish brown, the branchlets gray-pubescent; lf. blades lanceolate to elliptic or obovate, 1–2.5 cm long, mucronate, glabrous to somewhat puberulent at the base, petioles 3–7 mm long, pubescent; infl. compact, many fld., the bracts 2–3 mm long, pubescent; pedicels 2–5 mm long, pubescent to glabrous; sepals mostly white, ca. 1.5 mm long; corolla white, urceolate, 3–5 mm long; anthers bearing 2 hooked awns, filaments dilated at base, pubescent; fr. pulpy, depressed-globose, 5–6 mm in diam. Common. Open slopes and woods, 5,600–7,500 ft. Red Fir F., Subalpine F. Josephine L. Basin (*71*), Packers Pk. (*559, 936*), Le Roy Mine (*669*). June–August.

A. patula Greene. Green-leaved Manzanita. Fig. 36. Stout shrubs, 1–2 m tall, branches spreading from an enlarged rootcrown; bark smooth, reddish brown, the branchlets glandular-pubescent; lf. blades broadly ovate to nearly round, 1–4.5 cm long, glabrous to somewhat glandular at the base, petioles stout, 6–12 mm long, glandular-pubescent; infl. compact, many fld., the bracts mostly 2–4 mm long, glandular-pubescent; pedicels glabrous, 3–5 mm long; sepals pinkish, 1.5–2 mm long; corolla white to pinkish, urceolate, 5–8 mm long; anthers bearing 2 hooked awns, filaments dilated at base, pubescent; fr. pulpy, flattened-globose, 7–10 mm broad. Common. Dry open slopes, 5,000–7,500 ft. Montane Chaparral, Mixed Conifer F., Red Fir F. Big Flat (*492*), Dorleska Summit (*579*). June–July.

Cassiope D. Don

C. mertensiana (Bong.) G. Don. Western Mountain Heather. Fig. 37. Prostrate shrub with ascending branches to 3 dm high; the upper branches densely clothed with the scalelike lvs.; lvs. lanceolate to ovate, 2–4 mm long, closely imbricate, sessile, 4-ranked, ± keeled on the back; fls. solitary in the lf. axils toward the ends of the branches, nodding on long, slender pedicels; sepals pink to reddish, ca. 2 mm long; corolla white, campanulate, 5–6 mm long; anthers bearing slender awns; capsule rounded, ca. 3 mm long. Common. Damp, rocky slopes and ledges, 5,900–8,400 ft. Subalpine F., Alpine Fell-fields. Caribou Mt. (*351*), Caribou Basin (*733*), Canyon Cr. (*820*). July–August.

Kalmia L.

K. polifolia Wang. var. *microphylla* (Hook.) Rehd. Kalmia. Fig. 38. Low, spreading shrub, 1–2 dm high; bark brown to gray; lvs. oblong, obovate to narrowly elliptic, glabrous, ± revolute, 1–2 cm long, opposite; infl. a few fld. cyme; pedicels long and slender; sepals 2.5–3 mm long, rose; corolla rotate, 8–12 mm wide, 10-saccate at the base of the lobes, rose-purple; capsule globose, 5–6 mm in diam. Common. Wet meadows and slopes, 5,600–7,800 ft. Red Fir F., Subalpine F., Alpine Fell-fields. Boulder Cr. Basin (*178*), head

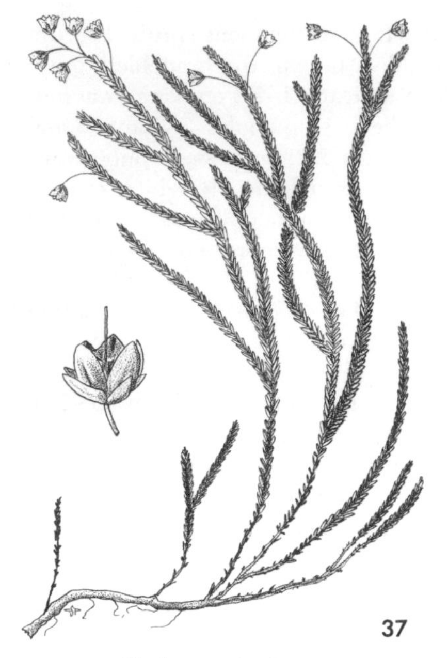

Cassiope mertensiana

Kalmia polifolia var. microphylla

Ledum glandulosum var. californicum

Phyllodoce empetriformis

of Canyon Cr. (*813*), below ice field on the north side of Thompson Pk. (*912*). July–August.

Ledum L.

L. glandulosum Nutt. var. *californicum* (Kell.) C. L. Hitchc. Labrador Tea. Fig. 39. Low shrub ca. 5 dm tall; bark brownish to gray; lvs. crowded at the ends of the branches, oblong to obovate, 1–3.5 cm long, glabrous and shining above, bearing resin granules and a minute puberulence below; infl. several fld., glandular-pubescent; sepals 1.5–2 mm long, ciliate; petals white, quite separate to base, 5–8 mm long; anthers on long slender filaments that are pubescent at the base; capsule brownish, globose, 4–5 mm in diam., glandular-rugose, splitting from base. Occasional. Stream banks and open slopes, 5,600–6,800 ft. Red Fir F., Subalpine F. Caribou Basin (*759*), Canyon Cr. (*776*), Landers L. (*1016*). July–August.

Leucothoe D. Don

L. davisiae Torr. Western Leucothoe. Evergreen shrub 5–15 dm tall; lvs. oblong to elliptic, glabrous, serrulate, 1.5–6 cm long; infl. of several elongate, many fld. racemes; pedicels recurved, 3–7 mm long; sepals whitish, 2–3 mm long; corolla white, urceolate, 6–8 mm long; anthers mucronate; capsule depressed-globose, 4–5 mm across. Common. Damp meadows and slopes, 5,800–8,000 ft. Red Fir F., Subalpine F. Josephine L. (*279*), Caribou Mt. (*353*), Boulder Cr. L. (*473*), Caribou Basin (*739*), Canyon Cr. (*833*). July–August.

Phyllodoce Salisb.

P. empetriformis (Sm.) D. Don. Pink Mountain Heather. Fig. 40. Low spreading shrub, 1–3 dm high; bark brown to gray; lvs. linear, needlelike, 6–13 mm long, minutely glandular-serrulate; fls. several, borne in the axils of persistent bracts on long, slender glandular-pubescent pedicels; sepals purplish, ca. 2 mm long; corolla rose to purple, campanulate, 5–6 mm long; capsule globose, 3–4 mm in diam. Common. Damp, open slopes and edges of bogs, 6,000–8,400 ft. Red Fir F., Subalpine F., Alpine Fell-fields. Canyon Cr. (*164*), Caribou Basin (*204, 726*), Caribou Mt. (*352*). July–August.

Vaccinium L.

V. arbuscula (Gray) Merriam. Dwarf Huckleberry. Low shrub to 4 dm tall; branchlets glabrous to puberulent; lvs. obovate, prominently veined, 10–15 mm long, serrulate, each tooth bearing a gland-tipped hair; fls. solitary in axils, nodding; calyx lobes obscure; corolla white to pinkish, urceolate, 3–4 mm long; anthers bearing slender awns and each sac prolonged into a tubular appendage with an apical pore; berry blue-black, globose, 5–7 mm in diam. Occasional. Open, rocky places, 5,800–7,400 ft. Red Fir F., Subalpine F. Caribou Basin (*205*), Boulder Cr. L. (*462*). July–August.

EUPHORBIACEAE

Euphorbia L.

Cyathium bearing white, petaloid appendages; glands not bearing horns . . . . *E. serpyllifolia*
Cyathium lacking petaloid appendages; glands bearing 2 slender horns. *E. crenulata*

E. crenulata Englm. Chinese Caps. Fig. 41. Glabrous annual or biennial; stems erect 10–15 cm tall; cauline lvs. obovate to spatulate, entire, 5–12 mm long, floral lvs. deltoid, ca. 10 mm long; cyathium ca. 2 mm long, the lobes fimbriate; glands ca. 1 mm long with 2 slender horns; petaloid appendages lacking. Found at one location. Dry slope, 5,000 ft. Montane Chaparral. Yellow Rose Mine Trail (*568*). June–July.

E. serpyllifolia Pers. Thyme-leaved Spurge. Procumbent annual; stems glabrous, several from a slender taproot, 1–3 cm long; lvs. opposite, ovate to oblong, apically toothed, 1–5 mm long; cyathium ca. 1 mm long, lobes minute, glands transversely oblong, ca. 0.3 mm wide subtended by white petaloid appendages. Found at one location. Dry lake bed, 5,000 ft. Mixed Conifer F. Big Flat (*361*). August.

FAGACEAE

Fr. 1–3 nuts enclosed in a spiny bur. *Castanopsis*
Fr. an acorn, the nut in a cuplike involucre . . . . *Quercus*

Castanopsis Spach.

C. sempervirens (Kell.) Dudl. Bush Chinquapin. Low, spreading shrubs 3–10 dm high; lvs. narrowly elliptic, 2.5–6 cm long, dull green and glabrous above, covered with a dense golden-yellow tomentum below; petioles 3–8 mm long; male catkins 2–5 cm long, in clusters toward ends of branches; fr. of 1–3 thin shelled nuts enclosed in a spiny involucre forming a bur 2–3 cm wide. Occasional. Dry, rocky slopes, 5,000–7,000 ft. Mixed Conifer F., Montane Chaparral. Caribou Mt. (*1385*). July–August.

Quercus L.

Lvs. deeply cleft, 3.5–8 cm long; nut glabrous on inner surface of shell. *Q. garryana* var. *breweri*
Lvs. entire to sharply toothed, 1–4 cm long; nut tomentose on inner surface of shell . . . . *Q. vaccinifolia*

Q. garryana Dougl. var. *breweri* (Engelm. in Wats.) Jeps. Oregon Oak. Low, spreading shrubs to small trees, 1–3 m high; bark scaly, light gray; lvs. 3.5–8 cm long, 2–5 cm wide, deeply cleft, shiny green and sparingly pubescent above, dull and densely stellate-pubescent below; nut ovoid, rounded at apex, ca. 2.5 cm long, glabrous on the inner surface of the shell; cup 12–20 mm broad, covered with flat, pubescent scales. Common. Dry, open slopes and flats, 5,000–6,000 ft. Mixed Conifer F., Montane Chaparral. Yellow Rose Mine Tr. (*1482*). June–July.

Q. vaccinifolia Kell. Huckleberry Oak. Low, spreading shrub, 1–10 dm high; branches slender, tufted; bark gray-brown; lvs. lanceolate, ovate to oblong, 1–4 cm long, entire to sharply toothed, dull green above, grayish green below, mostly glabrous in age, the young lvs. stellate-pubescent on one or both surfaces; male fls. in slender, drooping catkins on the current season's growth; female fls. solitary, in involucres in the upper axils; nut round-ovoid, 1–1.5 cm long, tomentose on the inner surface; the cup 10–14 mm broad, pubescent within, the scales tomentose. Common. Dry, open slopes and ridges, 5,000–7,500 ft. Montane Chaparral, Mixed Conifer F., Red Fir F. Boulder Cr. Basin (*458*), Yellow Rose Mine (*671, 956*), Black Mt. (*1085*). June–July.

FUMARIACEAE

Dicentra Bernh.

Fls. several; outer petals with erect to spreading tips 2–4 mm long. *D. formosa*

Fls. 1–3; outer petals with strongly recurved tips 6–9 mm long. *D. pauciflora*

D. formosa (Andr.) Walp. Pacific Bleeding Heart. Scapose perennial from a horizontal rootstock or a mass of fibrous roots; stems 20–30 cm tall; lvs. basal, the blades 3–10 cm long, biternately compound, the ultimate divisions linear, petioles slender, 5–15 cm long; fls. several; sepals 2–4 mm long; corolla light purple to brownish, 12–16 mm long, the outer pair of petals with erect to slightly spreading tips 2–4 mm long. Common. Damp shaded places, 5,000–6,500 ft. Mixed Conifer F., Red Fir F. Near Browns Meadow (*218*), near head of the South Fork of the Salmon R. (*416*), Big Flat (*497*), Canyon Cr. (*826*). June–August.

D. pauciflora Wats. Few-flowered Bleeding Heart. Scapose perennial from deep-seated fascicles of tuberous roots; stems 6–20 cm tall; lvs. basal, the blades 3–5 cm long, biternately compound, the ultimate divisions linear, petioles slender, 5–10 cm long; fls. 1–3; sepals 3–6 mm long; corolla pale purple to brownish, 15–25 mm long, the outer pair of petals with strongly recurved tips 6–9 mm long. Common. Damp, gravelly meadows and slopes, 5,500–6,500 ft. Red Fir F. Yellow Rose Mine Trail (*566*), Yellow Rose Mine (*574*), head of Union Cr. (*653*), head of Sunrise Cr. (*663*). June–July.

GARRYACEAE

Garrya Dougl.

G. fremontii Torr. Fremont's Silk-tassel. Fig. 42. Spreading to erect shrubs, 1–2 m high; lvs. opposite, elliptic, mucronate, 2.5–6 cm long, leathery, shiny

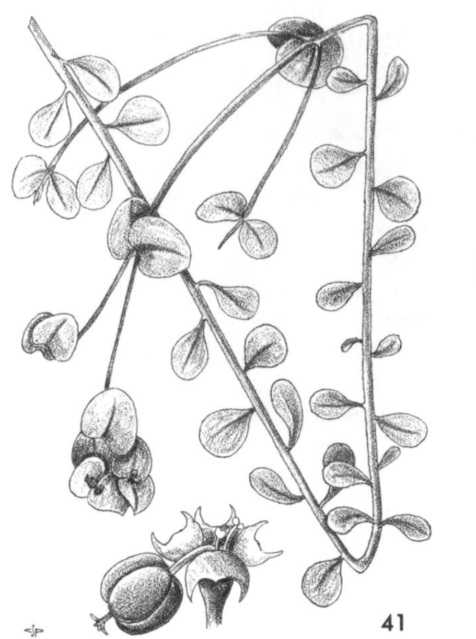

Euphorbia crenulata

Garrya fremontii

Gentiana calycosa

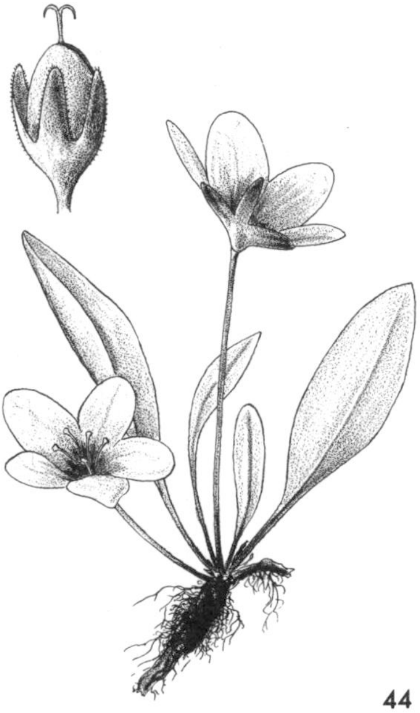

44

Hesperochiron pumilus

above, dull beneath, sparingly pubescent to glabrous; petioles 2–15 mm long; catkins 3–8 cm long, solitary or in groups of 2–3; fruiting catkins with 7–40 purple to black berries ca. 5 mm in diam. Common. Open, brushy slopes, 5,000–6,500 ft. Montane Chaparral. Big Flat (*1129, 1191*). June–July.

GENTIANACEAE

Gentiana L.

Fls. 5-parted; plants from a stout, woody caudex.
    Stems erect, 15–40 cm tall; corolla with greenish bands on the tube. *G. calycosa*
    Stems decumbent to ascending, 5–10 cm long; corolla with dark bands on the tube. . . . . *G. newberryi*
Fls. 4-parted; plants from slender rootstocks. . . . . *G. simplex*

G. calycosa Griseb. in Hook. Explorers' Gentian. Fig. 43. Perennial from a branched, woody caudex; stems several, erect, 15–40 cm tall; lvs. in 8–10 pairs, broadly ovate, 1–2.5 cm long, the lowermost reduced; fls. 1–3; calyx ca. 15 mm long, the lobes narrowed at the base; corolla deep blue with greenish bands on the tube, 3.5–5 cm long; capsule 1.5–2 cm long on a long stipe. Common. Meadows and stream banks, 6,000–7,000 ft. Red Fir F. Grizzly Cr. (*866*), ridge west of Bullards Basin (*961*), Landers L. (*1018*). August.

G. newberryi Gray. Alpine Gentian. Perennial from a stout caudex surmounting a taproot; stems decumbent to ascending, 5–10 cm long; basal lvs. spatulate to oblanceolate, 2–4 cm long, cauline lvs. in 3–7 pairs, oblanceolate, 1–2 cm long; fls. solitary at the ends of the stems; calyx 15–20 mm long with linear-oblanceolate lobes somewhat narrowed at the base; corolla deep blue with darker bands on the tube and green spots in the throat, 2.5–4 cm long; capsule 1–1.2 cm long on a shorter stipe. Common. Damp meadows and rocky seeps, 5,000–7,500 ft. Red Fir F., Subalpine F. Caribou Basin (*388*), Boulder Cr. Basin (*482*), Ward L. (*1092*). August.

G. simplex Gray. Hikers' Gentian. Annual or biennial from a slender rootstock; stems slender, erect, 10–22 cm tall; lvs. in 2–5 pairs, oblanceolate, 12–20 mm long, the lower ± clasping; fls. solitary, mostly 4-parted; calyx 12–20 mm long with lanceolate lobes; corolla greenish on the tube, blue in the throat and on the lobes, 2–3 cm long; capsule ca. 1.5 cm long on a long stipe. Occasional. Bogs and wet meadows, 5,000–6,500 ft. Mixed Conifer F., Red Fir F. Big Flat (*368*), head of Union Cr. (*1000*). August.

HYDROPHYLLACEAE

Fls. solitary to several, borne in the lf. axils.
    Corolla 3–6 mm broad; fls. from the axils of the cauline lvs. *Nemophila*

Corolla 12–18 mm broad; fls. from the axils of the basal rosette. *Hesperochiron*

Fls. many in congested, usually scorpioid cymes.

Style bifid at apex; lvs. pinnately divided into 7–11, toothed lfts. *Hydrophyllum*

Style cleft ca. half way; lvs. entire to lobed or having one pair of entire lfts. . . . . *Phacelia*

Hesperochiron Wats.

H. pumilus (Griseb.) Porter. Dwarf Hesperochiron. Fig. 44. Acaulescent perennial from a thickened vertical root; lvs. in a loose basal rosette, 2.5–4 cm long, oblanceolate to narrowly spatulate, gradually tapered to slender petioles; fls. 1–2 from the lf. axils; pedicels 2.5–4 cm long; sepals ca. 5 mm long; corolla rotate, 12–18 mm broad, white with purple veins, yellow and long-hairy in the throat, the lobes 5–6; stamens 5–6; capsule 5–9 mm long. Found at one location. Open, wet meadow, 5,000 ft. Mixed Conifer F. Big Flat (*507*). June–July.

Hydrophyllum L.

H. occidentale (Wats.) Gray. California Waterleaf. Fig. 45. Perennial from a tough rootstock often bearing fibrous roots; stems 5–25 cm tall pubescent; lvs. 5–15 cm long pinnate into 7–11 lfts., the lower separate, the upper becoming confluent, each lft. with 3–6 broad teeth along the lower margin; fls. many in a dense, globose cyme; sepals lanceolate, ca. 3 mm long; corolla whitish to purple, 6–10 mm long; stamens well exserted; style bifid only at the tip; capsule ca. 4 mm in diam.; seeds 1–2. Common. Shaded openings in woods and meadows, 5,000–6,500 ft. Mixed Conifer F., Red Fir F. Along the South Fork of the Salmon R. (*90*), north side of Caribou Mt. (*215*), head of Sunrise Cr. (*665*). June–July.

Nemophila Nutt. ex Barton

Auricles ca. as long as sepals; pedicels 3–6 mm long. . . . *N. pedunculata*

Auricles much shorter than sepals or absent; pedicels 6–16 mm long. *N. parviflora* var. *austinae*

N. parviflora Dougl. ex Benth. var. *austinae* (Eastw.) Brand. Small-flowered Nemophila. Low annual; stems weak, hispidulous, 5–10 cm high; lvs. opposite, 1–2.5 cm long, 3–13 mm broad, the lowest pair entire the others pinnately divided into deltoid lobes; fls. several, axillary; pedicels 6–16 mm long, recurved in age; sepals 2–3 mm long, the sinuses bearing much shorter auricles or those absent; corolla white with a tinge of blue, ca. 3 mm broad. Found at one location. Shaded, rocky meadow, 6,500 ft. Red Fir F. Near the head of Kidd Cr. (*698*). July.

N. pedunculata Dougl. ex Benth. Meadow Nemophila. Low annual; stems sparingly hispid, 3–5 cm high; lvs. opposite, 1–2.5 cm long, 3–5 mm wide, the lowest pair entire, the others pinnately divided into linear lobes; fls. several, axillary; pedicels 3–6 mm long, ascending; sepals 1–2 mm long, the sinuses bearing strongly reflexed auricles ca. as long; corolla white to bluish with darker spots, ca. 5 mm broad. Found at one location. Damp, open meadow, 5,000 ft. Mixed Conifer F. Big Flat (*503*). June.

Phacelia Juss.

Stems 6.5–9 dm tall; lvs. all deeply lobed . . . . *P. procera*

Stems 1–4 dm tall; lvs. mostly entire, a few with a basal pair of lobes.

Plants from a heavy, woody caudex; lvs. gradually tapering to petioles . . . . *P. corymbosa*

Plants from a slender caudex; lvs. abruptly contracted to petioles. *P. mutabilis*

P. corymbosa Jeps. Serpentine Phacelia. Perennial from a stout, woody caudex covered with the remains of old lvs.; stems 1–several, ± glandular-pubescent and with longer, stiffer hairs, 1–2.5 dm tall; lvs. mostly entire, a few with basal lobes, silvery-strigose, linear-oblanceolate, 2–6 cm long, gradually tapered to petioles ca. as long, basal lvs. crowded, the cauline few and reduced; fls. many in several dense, scorpioid cymes ± crowded toward the ends of the stems; sepals linear-oblanceolate, hirsute, 7–9 mm long; corolla white, campanulate, 5–6 mm long; stamens well exserted, ca. 10 mm long; style cleft ca. half way; capsule hirsute, ca. 4 mm long; seeds 1–2. Occasional. Dry, open, rocky ridges, 7,000–7,800 ft. Red Fir F., Subalpine F. Dorleska Summit (*306*), Red Rock Mt. (*1023*). July–August.

P. mutabilis Greene. Changeable Phacelia. Fig. 46. Perennial from a slender, simple to branched caudex; stems 1–several, hirsute, 1–4 dm tall; lvs. entire to having a pair of lobes at the base of the blade, hirsute, the blades lanceolate to ovate, 2–5 cm long, abruptly contracted to the long, slender petioles, basal lvs. crowded, the cauline few and reduced; fls. many, in several dense, scorpioid cymes, ± scattered along the stem; sepals linear-oblanceolate, hirsute, 3–6 mm long; corolla white to purple, campanulate, 5–6 mm long; stamens well exserted, 7–9 mm long; styles cleft ca. half way; capsule stiff pubescent, ca. 3 mm long; seeds 1–4. Common. Dry, rocky places, 5,000–6,000 ft. Mixed Conifer F., Red Fir F. Big Flat (*146*), Canyon Cr. (*180*), Caribou Basin (*206*), near head of the South Fork of the Salmon R. (*418*), Grizzly Cr. (*872*). June–August.

P. procera Gray. Tall Phacelia. Perennial from a heavy root crown; stems simple, puberulent, becoming glandular in infl., 6.5–9 dm tall; lvs. well distributed on the stem, pubescent, deeply lobed, ovate, 4–9 cm long, abruptly contracted to slender petioles ca. half as long; fls. many in several scorpioid cymes crowded at the end of the stem; sepals linear-oblanceolate, glandular-pubescent, 5–7 mm long; corolla white, campanulate, ca. 6 mm long; stamens ca. 10 mm long, well exserted; style cleft ca. half way; capsule glandular-pubescent, 6–7 mm long; seeds 12–16. Occasional. Open, rocky slopes, 5,500–6,500 ft. Red Fir F. Bullards Basin (*635*), Canyon Cr. (*786*). July–August.

HYPERICACEAE

Hypericum L.

Stems 4–10 cm tall; lvs. 3–6 mm long. . . . . *H. anagalloides*
Stems 30–60 cm tall; lvs. 7–20 mm long. . . . . *H. perforatum*

H. anagalloides Cham. & Schlecht. Tinker's Penny. Low annual; stems slender, 4–10 cm long; lvs. ovate, 3–6 mm long; fls. solitary at the ends of the stems; sepals ca. 2 mm long; petals golden yellow, ca. 2.5 mm long; capsule ca. 3 mm long. Occasional. Wet meadows, 5,000–7,000 ft. Mixed Conifer F., Red Fir F. Big Flat (*142*). June–July.

H. perforatum L. Klamath Weed. Perennial from a slender caudex; stems tough, wiry, 30–60 cm tall; lvs. ovate, 7–20 mm long, the lower ones bearing sterile axillary shoots; fls. several in terminal cymes; sepals ca. 2 mm long; petals yellow with black dots on the margins; capsule 7–8 mm long. Found at one location. Wet meadow, 5,000 ft. Mixed Conifer F. Big Flat (*224*). July–August.

LABIATAE

Corolla nearly equally lobed.
  Fls. in dense, globose heads . . . . *Monardella*
  Fls. in axillary cymes . . . . *Trichostema*
Corolla unequally lobed and distinctly 2-lipped.
  Calyx with a helmetlike crest on the back. . . . . *Scutellaria*
  Calyx lacking such a crest.
    Upper pair of stamens longer than the lower pair, deflexed downward . . . . *Agastache*
    Upper pair of stamens shorter than or equal to the lower pair.
      Corolla tube with an oblique ring of hairs on the inside toward the base . . . . *Stachys*
      Corolla lacking such a ring . . . . *Prunella*

Agastache Clayt.

A. urticifolia (Benth.) Kuntze. Nettle-leaved Horse-mint. Perennial from a horizontal rootstock; stems mostly 5–10 dm tall, simple to branched above, puberulent; lvs. in several remote pairs, glabrous above, ± puberulent below, coarsely toothed, ovate, 3–8 cm long, contracted to slender, short petioles; fls. many in dense whorls which are crowded into a continuous terminal spike; calyx nearly regular, 8–9 mm long, strongly nerved, glandular-pubescent, with 5 purple, lanceolate teeth; corolla whitish to rose, 12–15 mm long; stamens exserted, the upper pair declined and longer than the lower pair; nutlets 1.5–2 mm long. Common. Damp, shaded places, 5,000–6,200 ft. Mixed Conifer F., Red Fir F. Big Flat (*258*), near head of the South Fork of the Salmon R. (*425*), Kidd Cr. (*440*). July–August.

Monardella Benth.

M. odoratissima Benth. ssp. *pallida* (Heller) Epl. Mountain Monardella. Fig. 47. Aromatic perennial from a branched woody rootstock; stems several, 10–30 cm tall, puberulent; lvs. ovate to lanceolate, entire, 1–3 cm long, glandular-punctate and ± puberulent on both surfaces, ± abruptly contracted to narrowly winged petioles 3–7 mm long; fls. many in dense globose heads; the 2 outer bracts often lf.-like and strongly reflexed, the others thin, often purplish, pubescent and ciliate; calyx green to pinkish or purplish, strongly nerved, woolly pubescent, especially about the teeth, often resinous dotted, 5–7 mm long; corolla lavender to pale pinkish purple, 10–15 mm long with subequal linear lobes, the tube exserted from the calyx and pubescent; stamens exserted, glabrous. Common. Dry, rocky slopes and ridges, 5,000–8,000 ft. Mixed Conifer F., Red Fir F., Subalpine F. Big Flat (*246*), Josephine L. Basin (*286*), Caribou Mt. (*339*), Packers Pk. (*933*), Red Rock Mt. (*1025*). August.

Prunella L.

P. vulgaris L. ssp. *lanceolata* (Barton) Hult. Self-heal. Perennial from slender, ± creeping rootstocks; stems 5–20 cm tall, pilose, especially above and at the nodes; lvs. ovate to lanceolate, entire to crenulate, 2–4.5 cm long, contracted to narrow petioles ca. as long; fls. many in a dense terminal spike with broad, ± ciliate bracts; calyx spine-tipped, ca. 8–9 mm long, 2-lipped, the upper ± entire, the lower lip with 2 lanceolate lobes; corolla purple, ca. 10 mm long; stamens included in the throat, the upper pair much shorter than the lower pair; nutlets ca. 2 mm long. Found at one location. Wet meadow, 5,000 ft. Mixed Conifer F. Big Flat (*143*, *370*). June–August.

Scutellaria L.

Corolla blue . . . . *S. antirrhinoides*
Corolla white . . . . *S. californica*

S. antirrhinoides Benth. Snapdragon Skullcap. Fig. 48. Perennial from slender rootstocks; stems pubescent with ± upcurled hairs, 10–25 cm tall; lvs. entire, oblong-ovate to lanceolate, 1–2 cm long, puberulent with curled hairs, sessile or on petioles 2–15 mm long; fls. several, in the upper axils; calyx 3–5 mm long, puberulent and ± glandular; corolla blue with white markings, ca. 15 mm long, pubescent; stamens included in the galeate upper lip. Common. Dry slopes and meadows, 5,000–6,700 ft. Mixed Conifer F., Red Fir F. Big Flat (*378*, *674*), along the South Fork of the Salmon R. (*266*), ridge west of Bullards Basin (*959*). August.

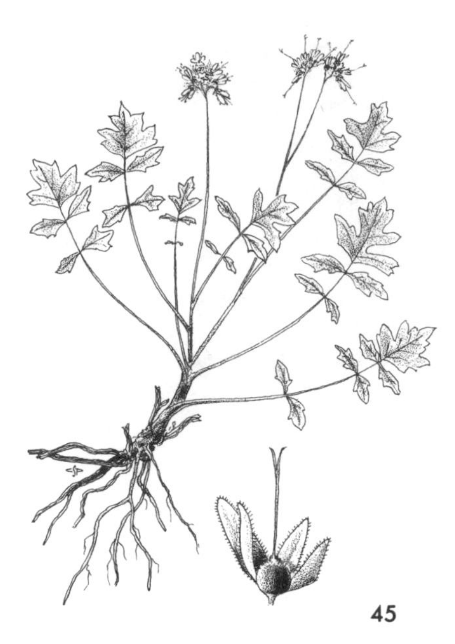

Hydrophyllum occidentale

Phacelia mutabilis

Monardella odoratissima ssp. pallida

Scutellaria antirrhinoides

S. californica Gray. California Skullcap. Perennial from a slender rootstock; stems 15–25 cm tall, puberulent with upcurved hairs and glandular; lvs. entire to crenate below, ovate to lanceolate, 1–2.5 cm long, puberulent with curled hairs and glandular, on slender petioles 2–5 mm long; fls. several, in the upper axils; calyx 4–5 mm long, glandular-puberulent; corolla white, tinged yellow, glandular-pubescent, 12–15 mm long; stamens included in the galeate upper lip. Found at one location. Dry, rocky slope, 5,400 ft. Montane Chaparral. Canyon Cr. (*841*). August.

Stachys L.

S. rigida Nutt. ex Benth. Rigid Hedge Nettle. Perennial from a slender rootstock; stems simple to branched above, hirsute, 4–7 dm tall; lvs. lanceolate to oblong-ovate, crenate-serrate, thinly to densely hirsute, 4–10.5 cm long, 1.5–5.5 cm wide, lowest petioles 1–2.5 cm long; fls. many, the whorls in elongate, interrupted spikes; calyx 5–6 mm long with 5 triangular-lanceolate spinose teeth; corolla whitish to lavender with purple lines and spots, the tube exserted, 7–8 mm long with an oblique ring of hairs near the base indicated by a constriction, the upper lip ca. 3 mm long, the lower lip 5–6 mm long; stamens exserted from the tube 2–3 mm. Common. Damp, shaded places, 5,000–6,200 ft. Mixed Conifer F., Red Fir F. Big Flat (*242*), Josephine L. (*288*), Kidd Cr. (*439*), near head of the South Fork of the Salmon R. (*414*), Canyon Cr. (*850*). August. These plants are extremely variable as to lf. size and pubescence.

Trichostema L.

T. oblongum Benth. Mountain Blue-curls. Slender stemmed, soft-villous annual; stems simple to branched below, 6–15 cm tall; lvs. oblanceolate, 1–2 cm long; fls. many, in several axillary cymes; calyx glandular, ca. 3 mm long, the narrowly lanceolate lobes twice as long as the tube; corolla blue to purplish, ca. 6 mm long with 5 nearly equal lobes, the tubes exserted from the calyx; stamens 6–7 mm long, well exserted and arched upward; nutlets 1–2 mm long. Found at one location. Dry edge of meadow, 5,000 ft. Mixed Conifer F. Big Flat (*367*, *380*). August.

LEGUMINOSAE

Lvs. palmately divided.
- Lfts. 5–8; fls. in elongate racemes .... *Lupinus*
- Lfts. 3; fls. in dense heads.......................... *Trifolium*

Lvs. pinnately divided.
- Fls. in umbels or solitary in lf. axils...................... *Lotus*
- Fls. in racemes.................................. *Astragalus*

Astragalus L.

A. whitneyi Gray var. *siskiyouensis* (Rydb.) Barneby. Whitney's Locoweed. Perennial from a branched, underground caudex; stems 1.5–3 dm tall, decumbent to ascending, thinly pubescent with appressed hairs; lvs. pinnately compound, lfts. 7–15, lanceolate, 1–2 cm long, sparsely pubescent with appressed hairs above, densely so below, the lower stipules united around the stem opposite the petiole; fls. 5–10 in an open raceme; pedicels ± recurved in fr., 1–2 mm long; calyx 4–6 mm long, thinly pubescent; corolla yellowish white; pods inflated, papery, glabrous, mottled with purple, 3–4 cm long, tapering to a slender stipe. Found at one location. Dry, open, rocky slope, 6,800 ft. Red Fir F. Packers Pk. (*934*). July–August.

Lotus L.

Stipules membranous, well developed.
  Peduncles bearing a 1–5 foliolate bract remote from the fls.; corolla cream with red markings . . . . *L. crassifolius*
  Peduncles bearing a 1–3 foliolate bract immediately below the fls.; corolla yellow . . . . *L. oblongifolius*
Stipules reduced to reddish, glandlike dots.
  Fls. solitary in the lf. axils . . . . *L. purshianus*
  Fls. borne in axillary umbels . . . . *L. nevadensis*

L. crassifolius (Benth.) Greene. Broad-leaved Lotus. Perennial from a branched, woody caudex; stems several, glabrous, glaucous, 5–6 dm tall; lvs. pinnately compound, lfts. 7–13, obovate to elliptic, mucronate, 1–3 cm long, pubescent with curled hairs, becoming glabrate in age; stipules membranous; fls. 6–13 in an umbel; peduncles 3–6 cm long with a 1–5 foliolate bract remote from the umbel; calyx ca. 6 mm long with short, subulate teeth; corolla cream with red in the center of the petals, 10–12 mm long, claws exserted from the calyx; pods 3.5–6.5 cm long. Found at one location. Dry woods, 5,500 ft. Mixed Conifer F. Kidd Cr. Trail (*701*). July.

L. cf. *nevadensis* Greene. Sierra Nevada Lotus. Perennial from a branched caudex; stems decumbent to ascending, wiry, many, ± forming a mat, pubescent with upcurled hairs, 1.5–2.5 dm long; lvs. pinnately compound, lfts. 3–5, obovate, 3–8 mm long, pubescent with appressed, curled hairs; stipules reduced to reddish glandular dots; fls. many in 4–7 fld. axillary umbels; peduncles 2–5 mm long; bract mostly 1-foliolate; calyx 2–3 mm long with subulate teeth ca. half as long as the tube; corolla yellow often tinged reddish in age, 5–6 mm long; pods strongly curved, ca. 6 mm long with a long, slender beak. Found at one location. Dry, open woods, 5,000 ft. Mixed Conifer F. Big Flat (*711*). July. *L. nevadensis* is described as having mat-forming stems 1.5–4 dm long, lfts. 7–15 mm long, lower peduncles 10–25 mm long; in California it has been previously reported only from the Sierra Nevada and southern mts. (Munz, 1959).

L. oblongifolius (Benth.) Greene. Narrow-leaved Lotus. Perennial from a simple to branched caudex; stems several, glabrate to pubescent, 2–4 dm tall;

lvs. pinnately compound, lfts. 5–7, oblanceolate, glabrate to pubescent, 1–2.5 cm long; stipules membranous; fls. 5–7 in an umbel immediately subtended by a 1–3 foliolate bract; peduncles 8–15 cm long in age; calyx glabrate to pubescent, 4–5 mm long with lance-linear teeth; corolla yellow, ca. 10 mm long, the claws barely exserted; pods 3–4.5 cm long. Occasional. Wet to drying places, 5,000 ft. Mixed Conifer F. Near head of the South Fork of the Salmon R. (*412*), Big Flat (*710*). August. In California this species has been previously reported from Monterey Co. to southern California and the Sierra Nevada (Munz, 1959).

L. purshianus (Benth.) Clem. & Clem. Spanish Clover. Thinly villous annual; stems simple to branched, 0.6–1.5 dm tall; lvs. pinnately compound; lfts. 3, obovate to elliptic, 5–10 cm long; stipules reduced to reddish glandlike dots; fls. several, solitary in lf. axils; peduncles 10–15 mm long; bract 1-foliolate; calyx 5–6 mm long with lance-linear teeth much longer than the tube; corolla white to cream, tinged red or lavender, 4–5 mm long; pods 1.5–2 cm long, deflexed. Common. Dry meadows and slopes, 5,000–6,000 ft. Mixed Conifer F. Big Flat (*259, 379*), Kidd Cr. (*682*). July–August.

Lupinus L.

Corollas yellow . . . . *L. croceus*
Corollas blue.
  Plants 3–7 dm high; pods 2–4 cm long.
    Lvs. pubescent on upper surface; keel glabrous . . *L. albicaulis*
    Lvs. glabrous on upper surface; keel ciliate on upper edges toward base . . . . *L. latifolius*
  Plants 0.7–1 dm high; pods 1–1.5 cm long . . . . *L. lyallii*

L. albicaulis Dougl. ex Hook. Sickle-keeled Lupine. Perennial from a branched, woody caudex; stems erect, puberulent, 4–7 dm tall; lfts. 2–5 cm long, oblanceolate, pubescent with appressed hairs on both surfaces, mostly mucronate; fls. many in an elongate raceme; calyx silky, ca. 6 mm long; corolla blue, 8–9 mm long, wings narrow and exposing the keel, the banner somewhat narrow; pods 3–4 cm long, silky-villous. Common. Open meadows, 6,000–7,200 ft. Red Fir F. Head of South Fork of the Salmon R. (*426*), Browns Meadow (*760*), Canyon Cr. (*808*), Ward L. (*1088*). July–August.

L. croceus Eastw. Mount Eddy Lupine. Fig. 49. Stout perennial from a branched, woody base; stems erect to spreading, ± strigose, 3–5 dm tall; lfts. oblanceolate, thinly silky-pubescent on both surfaces, 2–6 cm long, mucronate, petioles 3–6 cm long; fls. many in an elongate raceme; calyx silky, ca. 8 mm long; corolla yellow, 12–15 mm long, the wings mostly concealing the keel, the banner broadly ovate; pods 2–3.5 cm long, densely long-silky. Common. Dry woods, slopes, and ridges, 5,500–7,500 ft. Mixed Conifer F., Red Fir F., Subalpine F. Yellow Rose Mine (*297*), north side of Red Rock Mt. (*639*),

Kidd Cr. (*700*), Red Rock Mt. (*1024*). July–August.

L. latifolius J. G. Agardh. Broad-leaved Lupine. Erect perennial from a branched caudex; stems glabrous below, thinly strigose above, 3–6 dm tall; lfts. oblanceolate, glabrous on upper surface, thinly strigose below, 2.5–6 cm long, acute, petiole 1–2 cm long; fls. many in an elongate raceme; calyx villous. 6–7 mm long; corolla blue, ca. 10 mm long, the keel somewhat exposed, ciliate on the upper edges toward the base, the banner ovate; pods ca. 2 cm long, silky-pubescent with long hairs. Found at one location. Wet, open meadow, 7,000 ft. Red Fir F. Head of Kidd Cr. (*444*). August.

L. lyallii Gray. Lyall's Lupine. Low, densely clumped perennial from a branched rootcrown; stems 6–10 cm high; lvs. basal on petioles 3–4 cm long; lfts. oblanceolate, 8–15 mm long, strigose, the hairs 1–2 mm long; fls. several in racemes which are 3–4 cm long in fr.; calyx villous, ca. 5 mm long; corolla blue, 7–8 mm long, keel sharp-pointed, ciliate-villous on upper margins, mostly concealed by wings; pods villous, ca. 1 cm long. Rare. Dry, rocky to sandy slopes, 7,500–8,500 ft. Subalpine F. Ridge west of Thompson Pk. (*1301*), twin to Thompson Pk. (*Alexander & Kellogg 292*). July–August. In California this species has been previously reported only from the Sierra Nevada (Munz, 1959).

Trifolium L.

Heads subtended by an invol.
 Invol. bowl-shaped; calyx lobes trichotomously forked. ... *T. cyathiferum*
 Invol. flat, rotate; calyx lobes simple ... *T. wormskioldii*
Heads lacking an invol.
 Heads sessile, subtended by the uppermost lvs. ... *T. pratense*
 Heads borne on long, slender peduncles.
  Fls. pendulous; calyx glabrous ... *T. productum*
  Fls. erect to ascending; calyx villous ... *T. longipes*

T. cyathiferum Lindl. Bowl Clover. Low, glabrous annual 3–10 cm high; lfts. obovate, 3–10 mm long, spinulose denticulate; invol. bowl-shaped, finely toothed, 7–13 mm broad; fls. several to many; calyx 3–4 mm long, the teeth trichotomously forked; corolla pale yellowish green to pinkish. Found at one location. Open meadow, 5,000 ft. Mixed Conifer F. Big Flat (*510*). June–July.

T. longipes Nutt. Long-stalked Clover. Perennial with branching, creeping rootstocks from a central taproot; stems 5–15 cm tall, glabrous to thinly hirsute; lfts. linear to oblanceolate or elliptic, spinulose-serrulate, ± villous,

1–5 cm long, 4–10 mm broad, lfts. of the lower lvs. generally shorter and broader than those of the upper lvs.; fls. many, ascending to erect, sessile or short pediceled, in ovoid heads lacking an invol.; peduncles erect, 3–13 cm long; calyx villous with ascending hairs, 10–11 mm long, the subulate teeth much longer than the tube; corolla cream to purplish. Common. Seeps and bogs, 5,000–6,500 ft. Mixed Conifer F., Red Fir F. Yellow Rose Mine (*130*), Canyon Cr. (*156*), Boulder Cr. Basin (*174*), Big Flat (*513*), Dorleska Mine (*619*). June–July.

T. pratense L. Red Clover. Perennial from a slender taproot; stems several, sparsely pubescent below, becoming villous above, 10–20 cm tall; lfts. obovate, glabrous on upper surface, pubescent below, serrulate and ± ciliate, 1–2 cm long; fls. many in sessile heads immediately subtended by the uppermost lvs.; invol. lacking; calyx ca. 5 mm long, thinly villous, the subulate teeth slightly longer than the tube; corolla purple. Found at one location. Meadow, 5,000 ft. Mixed Conifer F. Big Flat (*257*). July–August.

T. productum Greene. Shasta Clover. Glabrous perennial from a stout taproot; stems several, 5–15 cm tall; lfts. mostly oblanceolate to obovate, spinose-serrulate, mucronate at the apex, 0.5–2.5 cm long; fls. many in a dense head, pendulous in age, the rachis of the infl. projected beyond the fls.; invol. lacking; peduncles 5–10 cm long; calyx purplish, 3–3.5 mm long, the subulate teeth ca. equal to the tube; corolla pale purple with darker purple lines. Occasional. Damp shaded to open slopes, 6,500–7,000 ft. Red Fir F. Kidd Cr. (*697*), Grizzly L. Basin (*888*). July–August.

T. wormskioldii Lehm. Cow Clover. Glabrous perennial from slender rootstocks; stems 10–20 cm tall; lfts. oblanceolate to elliptic, spinose-serrulate, mucronate, 0.5–2 cm long; fls. many in a rounded head subtended by a flattened, rotate, deeply divided invol.; peduncles 3–6 cm long; calyx 7–8 mm long, the subulate teeth somewhat longer than the tube; corolla purple. Found at one location. Open meadow, 5,000 ft. Mixed Conifer F. Big Flat (*228*). July–August.

LINACEAE

Linum L.

L. perenne L. ssp. *lewisii* (Pursh) Hult. Western Blue Flax. Fig. 50. Glabrous perennial from a tough caudex; stems several, slender, erect, 1.5–4 dm tall; lvs. many, sessile, linear to lance-linear, 1–2 cm long; fls. several in open racemes; pedicels slender, 1–2 cm long; sepals 3–5 mm long; petals blue, 1.5–2 cm long, falling very early; stamens 5; styles 5, stigmas capitate; capsule round-ovoid, acute at apex, 5–6 mm long. Occasional. Dry slopes, 6,500–7,000 ft. Red Fir F. Near Le Roy Mine (*668*), ridge west of Bullards Basin (*960*). June–August.

LORANTHACEAE

Phoradendron Nutt.

P. juniperinum Engelm. ssp. *libocedri* (Engelm.) Wiens. Libocedrus Mistletoe. Stems pendulous, brittle, 10–30 cm long; lvs. opposite, reduced to connate scales; male fls. 6 or 8; female spikes in pairs, 1-jointed, with 2 fls. per spike; sepals deltoid, ca. 1 mm long; berry sessile, globose, straw to wine colored. Common. Parasitic on *Calocedrus decurrens*, 5,000–6,000 ft. Mixed Conifer F. Big Flat (*1183, 1184*). June–July.

MALVACEAE

Fls. borne in axillary fascicles; lvs. obscurely lobed. . . . . *Malva*
Fls. borne in spikelike racemes; lvs. deeply lobed. . . . . *Sidalcea*

Malva L.

M. neglecta Wallr. Round-leaved Mallow. Procumbent biennial; stems 3–4 dm long, pubescent with forked hairs; lf. blades 1.5–3.5 cm broad, round-cordate, obscurely lobed, serrate, thinly pubescent with simple and stellate hairs; petioles 3–15 cm long, pubescent; fls. several, fascicled in the axils; pedicels 1–2.5 cm long; fls. subtended by an involucel of 3 linear bractlets; calyx 4–6 mm long, with acuminate lobes; petals white to pale pink, 7–10 mm long; stamens many, joined into a tube around the styles; styles several, linear, stigmatic on inner side; fruit depressed, disklike, separating into several 1-seeded carpels. Found at one location. Open weedy area, 5,000 ft. Mixed Conifer F. Big Flat (*261*). July–August.

Sidalcea Gray

S. oregana (Nutt.) Gray ssp. *spicata* (Regel) C. L. Hitchc. Oregon Sidalcea. Fig. 51. Perennial from a branched caudex; stems 4–6 dm tall, mostly hirsute with simple hairs at base; lvs. deeply divided, 3–10 cm broad, the cauline smaller and with narrower divisions; infl. a crowded spikelike raceme; pedicels 1–4 mm long; calyx 7–8 mm long with attenuate lobes, densely stellate-pubescent; petals pink to purple, 10–15 mm long; carpels ca. 3 mm long in fr. Common. Dry, rocky meadows, 5,000–6,700 ft. Mixed Conifer F., Red Fir F. Big Flat (*227*), Dorleska Mine (*311*), head of the South Fork of the Salmon R. (*419*), ridge west of Bullards Basin (*964*), head of Union Cr. (*1007*). July–August.

NYMPHAEACEAE

Nuphar Sm.

N. polysepalum Englm. Indian Pond Lily. Aquatic plants from stout, creeping rootstocks; lvs. floating or emergent, long petioled, the blades deeply cordate, 1–4 dm long; fls. usually standing above the water on a stout peduncle; sepals 5–12, yellow, rounded; petals many, narrow-cuneate, hidden by the many stamens; pistil several-celled with a disklike several-rayed stigma;

fr. ovoid to subglobose, 3–3.5 cm thick. Found at one location. Near edge of lake, 7,100 ft. Red Fir F. Ward L. (*456*). August.

ONAGRACEAE

Petals 2; capsule ca. 2 mm long, covered with hooked hairs. . . . *Circaea*
Petals 4; capsule longer, without hooked hairs.
  Seeds bearing a tuft of silky hairs at apex. . . . . *Epilobium*
  Seeds without a tuft of silky hairs at apex.
    Fls. inconspicuous, the petals white, turning purplish in age, 2–3 mm long. . . . . *Gayophytum*
    Fls. showy, the petals purple, 7 or more mm long . . . . *Clarkia*

Circaea L.

C. alpina L. ssp. *pacifica* (Asch. & Magnus) Raven. Enchanter's Nightshade. Delicate perennial from slender rootstocks; stems 10–30 cm tall; lvs. opposite, the blades ovate, rounded to cordate at base, thin, 1–7 cm long; petioles slender, 1–3 cm long; fls. many, in open, paniculately arranged racemes; sepals white to greenish, reflexed, 1.5–2 mm long; stamens 2; capsule ovate, ca. 2 mm long, covered with hooked hairs. Common. Damp, shaded places, 5,000–6,000 ft. Mixed Conifer F., Red Fir F. Josephine L. (*284*), Big Flat (*706*), Grizzly Cr. (*863*), along South Fork of the Salmon R. (*1034*). July–August.

Clarkia Pursh.

C. rhomboidea Dougl. Rhomboid Clarkia. Slender annual; stems simple, 10–20 cm tall, puberulent with upcurled hairs; lvs. few, lance-ovate to narrowly elliptic, 7–15 mm long, puberulent; fls. several, nodding in bud; sepals 6–8 mm long, puberulent; petals purple with darker spots, ca. 7 mm long, distinctly clawed with a pair of lateral projections toward the base; each stamen subtended by a white, ciliated scale ca. 1 mm long; capsule 1–2.5 cm long with a beak to 3 mm long. Occasional. Dry, rocky slopes, 5,500–6,000 ft. Montane Chaparral. Yellow Rose Mine Trail (*567*). June–July.

Epilobium L.

Petals 12–20 mm long.
  Plants erect, 6–15 dm tall. . . . . *E. angustifolium*
  Plants low, spreading, 0.5–1.5 dm tall. *E. obcordatum* ssp. *siskiyouense*
Petals 3–7 mm long.
  Plants of dry places; stems with exfoliating epidermis; greatly branched . . . . *E. paniculatum*
  Plants of damp to wet places; stems without exfoliating epidermis; simple or little branched.
    Stems 0.5–2 dm tall.

Rootstocks producing small, reddish winter buds.
E. pringleanum
Rootstocks not producing winter buds.
Capsule slender, 4–6 cm long; petals purple.
E. glaberrimum var. *fastigiatum*
Capsule ± clavate, 2–3 cm long; petals white to pinkish . . . . *E. clavatum*
Stems mostly 3–6 dm tall.
Rootstocks producing winter buds . . . . . . *E. brevistylum*
Rootstocks not producing winter buds.
Stems glabrous . . . . *E. glaberrimum*
Stems glandular-pubescent, at least above.
E. adenocaulon var. *occidentale*

E. adenocaulon Hausskn. var. *occidentale* Trel. Northern Willow-herb. Perennial from tough rootstocks; stems erect, 3–6 dm tall, ± branched and glandular-pubescent above; lvs. lanceolate, 1–4 dm long, serrulate, sparingly pubescent to glabrous, sessile to having short, broad petioles, mostly opposite; fls. many; pedicels up to 8 mm long in fr.; fl. tube 1.5–2 mm long; sepals ca. 2 mm long, puberulent; petals light purple, 5–6 mm long, notched at apex; capsule 4–6 cm long, pubescent, straight or somewhat curved. Found at one location. Bog, 5,000 ft. Mixed Conifer F. Big Flat (*226*). July–August.

E. angustifolium L. Fireweed. Stout perennial; stems 1–several, erect, 6–15 dm tall, simple, puberulent above, lvs. lance-linear, 4–12 cm long, entire to minutely denticulate, sessile to short petioled; fls. many; pedicels 7–8 mm long in fr.; fl. tube absent; sepals purplish, 12–14 mm long, puberulent; petals purple, 12–15 mm long, scarcely notched to entire at apex; capsule 5–7 cm long, canescent, straight. Common. Damp slopes and meadows, 5,000–7,500 ft. Mixed Conifer F., Red Fir F., Subalpine F. Big Flat (*245*), Sawtooth Ridge (*1079*). August.

E. brevistylum Barb. Slender Willow-herb. Perennial with compact winter buds at the base of the stem producing dried, persistent scales; stems mostly simple, 2–5 dm tall, glabrous to puberulent above; lvs. ovate to lanceolate, 1–5 cm long, glabrous to sparingly puberulent, sessile to short petioled; fls. few-several; pedicels ca. 5 mm long in fr.; fl. tube 1–1.5 mm long; sepals 2–2.5 mm long; petals purple, 5–7 mm long, notched at apex; capsule 3–4 cm long, ± glandular-pubescent. Occasional. Damp meadows and stream banks, 5,000–7,300 ft. Mixed Conifer F., Red Fir F., Subalpine F. Canyon Cr. (*849*), Caribou Basin (*1057*). August.

E. clavatum Trel. Clavate-fruited Willow-herb. Cespitose perennial; stems

Penstemon tracyi. This species is apparently endemic to the Trinity Alps and the Devils Canyon Mts., the type locality. It is shown here on a rock ledge at 6,500 ft. on Packers Pk. *P. tracyi* is common on Packers Pk. up to the summit but was not seen elsewhere in the study area. Plants are 10–15 cm high.

View south from the summit of the old Caribou Basin trail showing Caribou Lake, elev. 6,850 ft., and Sawtooth Mt., elev. 8,886 ft. Sawtooth Ridge is immediately in back of the lake. The broad, almost barren slopes on the far ridges below Sawtooth Mt. represent the ice carapaces of Sharp (1961). The steep cliffs which Sharp believes were formed by "bergschrunding" at the upper edges of these ice carapaces are also apparent. The contact between the Caribou Mt. Pluton and the Stuart Fork Formation can be seen running diagonally down Sawtooth Ridge into the southeast end of the lake.

Aerial view west along Sawtooth Ridge toward Thompson Pk. Caribou Lake is on the right, Emerald and Sapphire Lakes on the left. The southerly slopes of Sawtooth Ridge are densely covered with Montane Chaparral.

View south from Packers Pk. showing Red Rock Mt. (1), elev. 7,853 ft.; Black Mt. (2), elev. 8,019 ft.; and the head waters of the South Fork of the Salmon R. (3), elev. 5,500 ft. Caribou Mt. is at the upper right. The trees in the foreground are *Tsuga mertensiana* and *Abies magnifica* var. *shastensis*.

View south from Packers Pk. showing Thompson Pk. (1), elev. 9,002 ft., the highest point in the Trinty Alps, Sawtooth Mt. (2), and Caribou Mt. (3). The permanent ice fields on the north sides of Thompson Pk. and the unnamed peak just to the east, elev. 8,965 ft., each cover 5–6 acres. Photograph was taken in mid-July 1972.

Aerial view west over the South Fork of the Salmon River and the western Klamath region. Grouse Ridge is in the foreground with Caribou Mountain (1), elev. 8,575 ft., and Packers Pk. (2), elev. 7,828 ft. behind. Lower slopes are mainly forested with *Pinus jeffreyi*, *P. balfouriana*, *Pseudotsuga menziesii*, and *Abies magnifica* var. *shastensis*. Scattered patches of Montane Chaparral are also common.

View of Thompson Pk. from near Grizzly Lake. The lower edge of the ice field is at 7,800 ft. Trees on the lower slopes are mainly *Abies magnifica* var. *shastensis* and *Tsuga mertensiana*.

Aerial view south of Grizzly Lake, elev. 7,100 ft., and Thompson Pk., elev. 9,002 ft., immediately behind. The slope between the lake and the north face of Thompson Pk. has one of the shortest growing seasons in the Trinity Alps. *Claytonia nevadensis*, *Juncus drummondii*, *Luzula divaricata*, *Mimulus tilingii*, and *Saxifraga tolmiei* are among those species thriving here. Photograph was taken in late October, 1973.

Aerial view north of upper Canyon Cr. with Thompson Pk. at the upper left. Lower and Upper Canyon Cr. Lakes, at 5,600 ft. and 5,690 ft. respectively, are in the foreground. Many typically alpine species such as *Arnica mollis*, *Cassiope mertensianus*, *Hieracium gracile*, *Lewisia columbiana*, *Luetkea pectinata*, *Phyllodoce empetriformis*, and *Sibbaldia procumbens* are found on the upper slopes.

Aerial view east of Mt. Hilton, elev. 8,964 ft., in the center foreground and Sawtooth Mt. behind and to the right. Mt. Shasta is on the left horizon. On these higher ridges timberline is often near 8,000 ft. where stunted *Tsuga mertensiana* and *Pinus albicaulis* may be found.

View north from the summit of Black Mt. showing Packers Pk. (1), elev. 7,828 ft., and Big Flat (2), elev. 5,000 ft. The valley floor is forested mainly with *Pinus ponderosa*, *P. contorta* ssp. *latifolia*, *Abies concolor*, *Pseudotsuga menziesii*, and *Calocedrus decurrens*. Above 6,000 ft. on the adjacent slopes *Abies magnifica* var. *shastensis*, *Pinus jeffreyi*, and *P. balfouriana* are found. There are also extensive areas of chaparral on the slopes with *Ceanothus velutinus*, *Arctostaphylos patula*, and *Quercus vaccinifolia* being the most common species.

Aerial view northeast of the south side of Thompson Pk. (1), Packers Pk. (2) across the canyon of the South Fork of the Salmon River and the north flank of Caribou Mt. (3). Mt. Shasta is on the right horizon about 55 miles to the east. *Abies magnifica* var. *shastensis*, *Haplopappus lyallii*, *Hieracium gracile*, *Lupinus lyallii*, *Penstemon davidsonii*, *Pinus albicaulis*, *Polemonium pulcherrimum*, and *Tsuga mertensiana* are among the species found near the summit of Thompson Pk.

Aerial view east over Boulder Cr. Basin toward Sawtooth Mt., elev. 8,886 ft., with Mt. Shasta in the background. One of the Forbidden Lakes at 6,250 ft. is at the edge of the shadow in lower left. *Picea breweriana* is found in this basin.

One of the tallest (ca. 21 m) specimens of *Pinus balfouriana* seen in the study area. This tree and several others of similar proportions are found at the head of Union Cr., elev. 6,500 ft.

many, ± sigmoid, ca. 1 dm high, glabrous; lvs. obovate to elliptic, 0.5–1.5 cm long, denticulate, glabrous, short petioled; fls. few; pedicels 5–15 mm long in fr.; fl. tube ca. 1.5 mm long; sepals 3 mm long; petals white to pinkish, 5–6 mm long, notched at apex; capsule 2–3 cm long, ± clavate. Rare. Wet, open slopes, 7,800 ft. Alpine Fell-fields. Below ice fields on north side of Thompson Pk. (*917*). August.

E. glaberrimum Barb. Glaucous Willow-herb. Cespitose perennial from tough, branching rootstocks; stems several, mostly simple, 3.5–5.5 dm tall, glabrous; lvs. lanceolate, 1–3 cm long, denticulate, glabrous, green to purplish, mostly sessile with ± clasping bases, not crowded; fls. many; pedicels ca. 5 mm long in fr.; fl. tube ca. 1 mm long; sepals 3–4 mm long; petals purple, 5–6 mm long, notched at apex; capsule slender, 4–6 cm long. Found at one location. Damp gravel, 5,700 ft. Red Fir F. Canyon Cr. (*791*). July–August.

var. *fastigiatum* (Nutt.) Trel. Stems 0.5–1.5 dm tall; lvs. 0.5–2 cm long, mostly more crowded; fls. few-several. Found at one location. Damp, rocky slope, 7,000 ft. Red Fir F. Canyon Cr. (*818*). August.

E. obcordatum Gray ssp. *siskiyouense* Munz. Rock-fringe. Fig. 52. Low, cespitose, ± matted perennial, from a woody base; stems several, 0.5–1.5 dm long, mostly glabrous; lvs. ovate, 1–2 cm long, denticulate, glabrous to slightly puberulent, sessile with ± clasping bases to short petioled; fls. few; pedicels ca. 1 cm long in fr.; fl. tube 2–3 mm long; sepals 7–10 mm long; petals purple, 15–20 mm long, broadly notched at apex; capsule 2–3 cm long, glandular-puberulent. Common. Rocky ridges and slopes, mostly serpentine, 7,000 ft. Red Fir F., Subalpine F. Dorleska Summit (*303*), Red Rock Mt. (*973*), ridge between Landers L. and the head of Union Cr. (*1008*). August.

E. paniculatum Nutt. ex T. & G. Panicled Willow-herb. Erect, branched annual; stems 3–5 dm tall with many slender branches, glabrous; lvs. linear to linear-oblanceolate, 1–3 cm long, ± denticulate, glabrous, on short, slender petioles, often with axillary fascicles; fls. many; pedicels 5–12 mm long in fr.; fl. tube 1.5 mm long; sepals 2–3 mm long; petals light purple, 4–5 mm long, notched at apex; capsule 2–2.5 cm long, glandular-puberulent. Found at one location. Dry, sandy, open areas, 5,000 ft. Mixed Conifer F. Big Flat (*239*). July–August.

E. pringleanum Hausskn. Slender Willow-herb. Low perennial from slender rootstocks producing small, red, winter buds; stems simple, 1–2 dm tall, glabrate to pilose; lvs. linear-lanceolate, 1–2 cm long, denticulate, glabrous to pilose, sessile; fls. several; pedicels 3–5 mm long in fr.; fl. tube 1–1.5 mm long; sepals 2–3 mm long; petals purple, 5–6 mm long, notched at apex; capsule 3–4 cm long, ± glandular-pubescent. Common. Damp, open meadows, 5,000 ft. Mixed Conifer F. Big Flat (*95*, *505*). June–July.

Lupinus croceus

Linum perenne ssp. lewisii

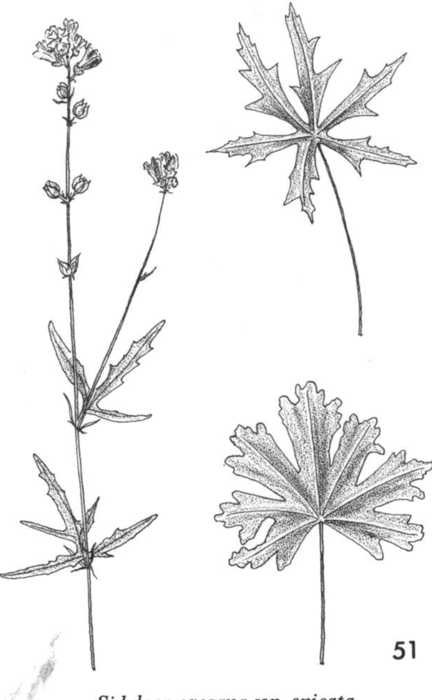

Sidalcea oregana ssp. spicata

Epilobium obcordatum ssp. siskiyouense

Gayophytum Juss.

G. diffusum T. & G. ssp. *parviflorum* Lewis & Szweyk. Nuttall's Gayophytum. Slender stemmed, diffusely branched annual, 20–40 cm tall; stems mostly glabrous; lvs. few, linear, 1–3 cm long; fls. many, inconspicuous, on slender pedicels 3–5 mm long; sepals 1–2 mm long; petals white, turning purplish in age, 2–2.5 mm long; stamens 8, the alternate set reduced; capsule 7–12 mm long, torulose, densely pubescent to nearly glabrous. Common. Dry slopes and meadows, 5,000 ft. Mixed Conifer F. Big Flat (*237*, *256*), head of South Fork of the Salmon R. (*413*), Canyon Cr. (*845*). August.

OROBANCHACEAE

Fls. many in a ± woody spike . . . . *Boschniakia*
Fls. solitary on a scapelike pedicel . . . . *Orobanche*

Boschniakia D. A. Mey.

B. strobilacea Gray. California Ground Cone. Parasitic perennial from a cormlike basal thickening at the junction with the host root; stems 10–15 cm tall; lvs. scalelike, ± imbricate, ca. 1 cm long; fls. many in a dense spike 3–4 cm thick; each fl. subtended by a lf.-like bract; corolla 10–15 mm long, dark purple, bent at the middle; capsule rounded, ca. 1 cm long. Found at one location. On *Arctostaphylos*, 7,200 ft. Montane Chaparral. Caribou Mt. (*345*). July–August.

Orobanche L.

O. uniflora L. var. *purpurea* (Heller) Achey. Naked Broomrape. Fig. 53. Glandular-pubescent root parasite; stems mostly underground with a few crowded bracts; fls. solitary on a scapelike pedicel, 3–5 cm long; calyx 6–7 mm long with lance-linear lobes; corolla blue-purple with yellow in the throat, 12–15 mm long; capsule ovoid, usually capped by the persistent dried corolla. Found at one location. Open meadow, 5,000 ft. Mixed Conifer F. Big Flat (*504*). June.

PLANTAGINACEAE

Plantago L.

P. major L. Common Plantain. Acaulescent perennial; lvs. in a basal rosette, the blades ovate, 2.5–4 cm long, several nerved, pubescent with short, stiff hairs, abruptly contracted to winged petioles; spikes narrow, cylindrical, curved-ascending, 3–4 cm long, bracts ovate, shorter than the calyx; sepals 1.5–2 mm long; corolla lobes 0.5 mm long; capsule broadly conic, 3–4 mm long, brown to purple. Found at one location. Dry lake bed, 5,000 ft. Mixed Conifer F. Big Flat (*383*). August.

POLEMONIACEAE

Calyx growing with the capsule and not ruptured by it, becoming chartaceous in age.
  Lvs. simple, entire; corolla 13–30 mm long. . . . . *Collomia*
  Lvs. pinnately compound; corolla 6–8 mm long. . . . . . *Polemonium*
Calyx not growing with the capsule, eventually ruptured by it.
  Lvs. entire.
    Corolla showy, 15–20 mm long with broad lobes. . . . . . *Phlox*
    Corolla inconspicuous, ca. 8 mm long with short, linear lobes. *Microsteris*
  Lvs. divided.
    Lvs. palmately divided, sometimes appearing whorled. *Linanthus*
    Lvs. pinnately divided.
      Lobes of lvs. and bracts rigidly spinose. . . . . . *Navarretia*
      Lobes of lvs. and bracts not spinose.
        Corollas bright red with yellow blotches. . *Ipomopsis*
        Corollas pale blue . . . . *Gilia*

Collomia Nutt.

Corolla salmon-yellow, 15–30 mm long. . . . . *C. grandiflora*
Corolla white to pink to bluish, 13–18 mm long.
  Stamens equally inserted; stigmas barely exserted . . . . *C. tinctoria*
  Stamens very unequally inserted; stigmas included. . . . . . *C. tracyi*

C. grandiflora Dougl. ex Lindl. Large-flowered Collomia. Slender, erect annual; stems simple to branched, 40–60 cm tall, reddish, pubescent above; lvs. linear-lanceolate, entire, sessile, 2–4 cm long, puberulent; fls. several in terminal and axillary heads, bracts glandular-pubescent; calyx 7–8 mm long, becoming chartaceous, glandular-pubescent; corolla 15–30 mm long, salmon-yellow; stamens unequally inserted, unequal in length; capsule obovoid, ca. 5 mm long. Occasional. Dry, open slopes and meadows, 5,000–6,000 ft. Mixed Conifer F., Red Fir F. Big Flat (*254*), Josephine L. Basin (*287*). July–August.

C. tinctoria Kell. Yellow-staining Collomia. Simple to branched annual; stems 4–15 cm tall, glandular-pubescent; lvs. opposite, mostly sessile, linear-oblanceolate, 1–2.5 cm long, ± glandular-pubescent; fls. many, 1–3 in axils or forks of branches, and in dense clusters at the ends of the branches, mostly short pediceled; calyx 5–7 mm long, becoming chartaceous in age, glandular-pubescent, the lobes lance-aristate; corolla white, turning pink in age, 13–18 mm long; stamens subequally inserted, unequal in length; stigmas barely exserted; capsule ca. 5 mm long. Found at one location. Dry, rocky slope, 6,000 ft. Red Fir F. Packers Pk. (*553, 944*). June–August.

C. tracyi Mason. Tracy's Collomia. Low annual with spreading branches; stems 5–15 cm tall, glandular-pubescent; lvs. opposite, petiolate, linear-oblanceolate, 1–4 cm long, ± glandular-pubescent; fls. several, mostly 1 or 2 in the axils or forks of the branches; calyx 5–6 mm long, glandular-puberulent, the lobes lance-aristate; corolla 12–15 mm long, white to pink; stamens very unequally inserted, unequal in length; stigmas included; capsule 4–5 mm long. Found at one location. Shaded stream bank, 6,400 ft. Red Fir F. Kidd Cr. (*1103*). August.

Gilia R. & P.

G. capitata Sims. Blue Field Gilia. Slender stemmed annual 10–20 cm tall; stems solitary and simple to several and branched, glandular-pubescent below, glabrous above; lvs. 1–3 cm long, pinnately divided into linear segments, becoming reduced above; fls. 10–25 in dense heads; calyx ca. 3 mm long, the lobes connected by a scarious membrane; corolla light blue, 6–8 mm long with linear lobes; capsule subglobose, ca. 3 mm long. Occasional. Dry, rocky slopes, 6,500–7,000 ft. Red Fir F. Northwest side of Caribou Mt. (*403*), Packers Pk. (*545*). June–August.

Ipomopsis Michx.

I. aggregata (Pursh) V. Grant. Scarlet Gilia. Fig. 54. Perennial from a somewhat woody caudex surmounting a tough taproot; stems few-several, spreading to erect, 3–6 dm tall; lvs. 2–5 cm long, pinnately divided into linear segments, sparingly tomentose; fls. many in open, elongate panicles; calyx 5–7 mm long with subulate lobes; corolla bright red with yellow blotches, tubular-funnelform, 2–3 cm long; capsule ovoid, 5–8 mm long. Common. Dry, open meadows and slopes, 5,000–7,000 ft. Mixed Conifer F., Red Fir F. Big Flat (*86*), north side of Caribou Mt. (*209*), Grizzly Cr. (*869*), head of Union Cr. (*990*). June–August.

Linanthus Benth.

Corolla tube well exserted from the calyx. . . . . *L. ciliatus*
Corolla tube included in the calyx.
  Corolla 3 mm long; slender stemmed annuals. . . . . *L. harknessii*
  Corolla 10–15 mm long; perennial from a branched, woody caudex. *L. nuttallii*

L. ciliatus (Benth.) Greene. Whisker-brush. Low, stiffly pubescent annual; stems simple to branched from base, 5–15 cm tall; lvs. opposite, palmately divided into several filiform segments; fls. sessile, several in a crowded, densely bracteate infl., the bracts similar to the lvs. and stiffly ciliate; calyx ca. 10 mm long, the filiform, acerose lobes united ca. half way by a hyaline membrane; corolla pink, turning darker in age, yellow in the throat, 15–20 mm long, the slender tube well exserted from the calyx; capsule oblong, ca. 6 mm long. Found at one location. Dry, open slope, 5,000 ft. Mixed Conifer F. Big Flat (*537*). June–July.

L. harknessii (Curran) Greene. Harkness' Linanthus. Slender stemmed annual 5–20 cm tall; lvs. opposite, palmately divided into several filiform segments; fls. several on long, slender pedicels in an open panicle; calyx ca. 2 mm long, the lobes joined by a hyaline membrane; corolla white, ca. 3 mm long, the tube included in the calyx; capsule ca. 2 mm long. Found at one location. Dry, open meadow, 5,000 ft. Mixed Conifer F. Big Flat (*604*). June–July.

L. nuttallii (Gray) Greene ex Mlkn. Nuttall's Linanthus. Perennial from a branched, woody caudex; stems several, 15–25 cm tall; lvs. opposite, palmately divided into several narrow segments; fls. sessile, several in crowded, bracteate heads; calyx 7–8 mm long, the narrow lobes partly joined by a narrow, hyaline membrane; corolla 10–15 mm long, the lobes cream, the tube pale yellow; capsule oblong, ca. 5 mm long. Found at one location. Dry, open slope, 6,500 ft. Red Fir F. Dorleska Mine (*308*). August.

Microsteris Greene

M. gracilis (Dougl. ex Hook.) Greene. Slender Phlox. Low annual; stems 5–10 cm high, simple to branched, ± pilose below, glandular-pubescent above; lvs. opposite below, sometimes alternate above, linear to linear-oblanceolate, 1–2 cm long; fls. several, from the axils of linear bracts, sessile or on pedicels to 10 mm long; calyx 6–7 mm long, the lobes joined ca. half way by a hyaline membrane; corolla ca. 8 mm long with short, linear, rose to lavender lobes, the tube yellowish; capsule ca. 5 mm long. Occasional. Dry slopes and meadows, 5,000 ft. Mixed Conifer F. Big Flat (*494, 515*). June–July.

Navarretia R. & P.

N. intertexta (Benth.) Hook. Needle-leaved Navarretia. Slender stemmed annual, 5–10 cm high; stem reddish brown, retrorse-pubescent with crisped white hairs; lvs. 1-2 cm long, pinnately divided into stiff, narrow, spinose segments; fls. several in dense, bracteate heads, the bracts foliose with stiff, spinose tips; calyx 7–8 mm long, the lobes unequal, simple to pinnately divided, spinose, ± pubescent on the tube and around the throat; corolla pale blue, ca. 8 mm long, the lobes 1.5–2 mm long; stamens exserted; stigmas 2-lobed, ca. 0.5 mm long. Found at one location. Dry, open meadow, 5,000 ft. Mixed Conifer F. Big Flat (*713*). July.

Phlox L.

P. diffusa Benth. Spreading Phlox. Fig. 55. Perennial from a branched, woody caudex, forming dense clumps; stems low, spreading, 5–15 cm long; lvs. opposite, crowded, 10–15 mm long, linear-subulate, acerose, ± pilose;

fls. several, showy; calyx 7–9 mm long, densely pilose, with narrow, acerose lobes; corolla pink to blue or purple, 15–20 mm long, the lobes broadly to narrowly obovate; capsule 4–5 mm long. Common. Dry, rocky slopes and ridges, 5,000–7,000 ft. Mixed Conifer F., Red Fir F. Canyon Cr. (*68*), Yellow Rose Mine (*128*), Big Flat (*536*), Dorleska Summit (*576*), trail from head of Union Cr. to Landers L. (*658*). June–July.

Polemonium L.

P. pulcherrimum Hook. Showy Polemonium. Perennial from a branched, woody caudex, forming dense clumps; stems ascending to erect, 5–15 cm tall, glandular-pubescent; lvs. mostly basal, 6–10 cm long, pinnately compound into 11–23 ovate to lanceolate lfts. 2–7 mm long; fls. several on slender pedicels 4–10 mm long; calyx 3–6 mm long, becoming chartaceous in age; corolla blue with a yellow throat, 6–8 mm long; capsule 3–4 mm long. Occasional. Cracks in rock cliffs and walls, 7,000–9,000 ft. Red Fir F., Subalpine F. Ridge between Landers L. and Union L. (*662*), Thompson Pk. (*823*). August.

POLYGALACEAE

Polygala L.

P. cornuta Kell. Sierra Milkwort. Fig. 56. Perennial from a branching woody base; stems 1–3 dm tall; lvs. ovate to linear-elliptic, 1–3 cm long, ciliolate; racemes 4–20 fld.; fls. cream, the outer 3 sepals ca. 4 mm long, densely puberulent, the inner 2 petaloid, puberulent, ca. 10 mm long; petals 3, the lower one forming a keel 11–12 mm long projecting into a slender beak; capsule ca. 8 mm long. Occasional. Dry slopes, 5,000–6,000 ft. Mixed Conifer F. Big Flat (*321*), Boulder Cr. Basin (*461*). August.

POLYGONACEAE

Lvs. with an evident stipular sheath.
  Calyx 4 or 6 parted.
    Calyx 4 parted; lvs. reniform; stigmas 2 . . . . *Oxyria*
    Calyx 6 parted; lvs. not reniform; stigmas 3 . . . . *Rumex*
  Calyx 5 parted . . . . *Polygonum*
Lvs. lacking such a sheath . . . . *Eriogonum*

Eriogonum Michx.

Calyx contracted to a narrow, stipelike base.
  Fls. whitish to pale yellow; lvs. 5–11 cm long . . . . . . *E. compositum*
  Fls. bright sulfur-yellow to reddish; lvs. 0.5–2.5 cm long.
    Primary rays of umbel branched or bearing bracts near the middle . . . . *E. umbellatum* var. *stellatum*
    Primary rays of umbel simple and lacking bracts near the middle . . . . *E. umbellatum*
Calyx not stipelike at base.
  Calyx glabrous externally . . . . *E. nudum*
  Calyx pubescent externally . . . . *E. elatum*

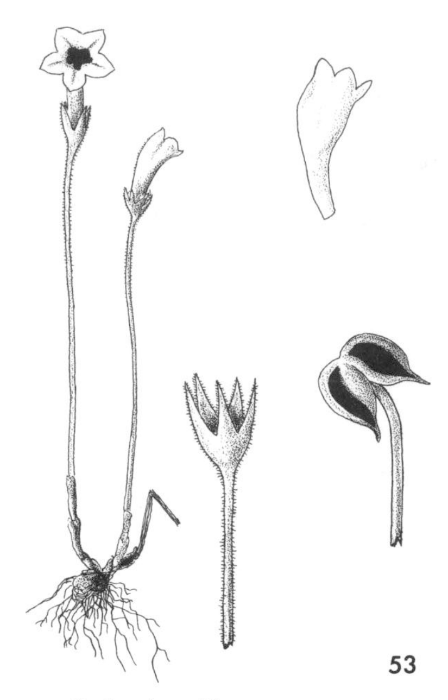

53

Orobanche uniflora var. purpurea

54

Ipomopsis aggregata

55

Phlox diffusa

56

Polygala cornuta

E. compositum Dougl. ex Benth. Composite Eriogonum. Fig. 57. Scapose perennial from a woody caudex; stems simple, glabrous to thinly pilose, 3–4 dm tall; lvs. basal, the blades lance-ovate, 4–8 cm long, greenish-glabrate above, white-tomentose below, abruptly contracted to slender, ± pilose petioles 5–11 cm long; fls. many in dense umbels subtended by linear foliaceous bracts 1–3 cm long; invol. campanulate, 1–3 mm long with reflexed lobes ca. half as long, tomentose; calyx 4–5 mm long, contracted to a narrow, stipelike base, pale yellow to whitish, glabrous; achenes light brown, pubescent above, ca. 5 mm long. Occasional. Open, rocky slopes, 6,000–7,000 ft. Red Fir F. Grizzly L. Basin (*884*). August.

E. elatum Dougl. ex Benth. Tall Eriogonum. Perennial from a branched, woody caudex; stems branched above, glabrous, 3–8 dm tall; lvs. basal, erect, narrowly ovate to lanceolate, 2–7 cm long, thinly villous on both surfaces, abruptly narrowed to slender, pilose petioles 1.5–7 cm long; fls. many in clusters at the ends of the branches; invol. glabrous, ca. 3 mm long, the teeth short, erect; calyx 2–3 mm long, white often tinged red, pubescent; achenes brownish, glabrous, ca. 4 mm long. Found at one location. Dry slope, 5,000 ft. Mixed Conifer F. Big Flat (*241*). July–August.

E. nudum Dougl. ex Benth. Naked-stemmed Eriogonum. Perennial from a branched, woody caudex; stems few-several, branched above, glabrous, 3–5 dm tall; lvs. basal, the blades oblong-ovate to ovate, 1–5 cm long, glabrate and green to reddish above, white-woolly lanate beneath, narrowed to slender, ± lanate petioles 3–7 cm long; fls. many in clusters at the ends of the branches; invol. cylindrical, 4–5 mm long with short erect teeth, glabrous; calyx ca. 3 mm long, white to pale yellowish, glabrous; achenes light brown, glabrous, ca. 3 mm long. Common. Dry, rocky slopes, 5,000–6,500 ft. Mixed Conifer F., Red Fir F. Canyon Cr. (*844*), Grizzly Cr. (*871*), head of Union Cr. (*988*), Big Flat (*1113*). August.

E. umbellatum Torr. Sulphur-flowered Eriogonum. Perennial from a branched, woody base; flowering stems thinly floccose, 1–2.5 dm tall; lvs. in tufts at the ends of the branches, oblanceolate to obovate, greenish-glabrate above, white-tomentose below, 5–25 mm long, narrowed to usually shorter petioles; primary rays of umbel simple, 2–5 cm long, the infl. subtended by 4–7 leafy bracts; invol. ca. 2 mm long with reflexed lobes, tomentose; calyx contracted to a narrow stipelike base, bright yellow to reddish, 6–7 mm long; achenes light brown, pubescent above, ca. 3.5 mm long. Common. Dry slopes, 5,000–7,500 ft. Mixed Conifer F., Red Fir F., Subalpine F. Big Flat (*240*), Packers Pk. (*935*). August.

var. *stellatum* (Benth.) Jones. Perennial with a branched, woody base; flowering stems floccose, 1.5–2 dm tall; lvs. in tufts at the ends of the branches, ovate to oblanceolate, greenish-glabrate above, densely white-tomentose below, 0.5–2 cm long, narrowed to usually shorter petioles; primary rays of the umbel 3–6 cm long, branched or bearing bracts near the middle, the infl. immediately subtended by 3–5 leafy bracts; invol. 3–4 mm long, with reflexed lobes, tomentose; calyx contracted to a narrow, stipelike base, bright sulfur-yellow to reddish, 5–6 mm long. Common. Dry, open slopes, 5,500–8,000 ft. Mixed Conifer F. Above head of the South Fork of the Salmon R. (*424*), Black Mt. (*1083*). August.

Oxyria Hill.

O. digyna (L.) Hill. Mountain Sorrel. Fig. 58. Perennial from a stout, woody, branched caudex; stems slender, erect, 8–30 cm tall; lvs. basal, round-reniform, 1.5–3 cm broad, petioles slender, 3–7 cm long; fls. many in crowded, elongate panicles; pedicels slender, 2–4 mm long, ± reflexed; sepals 4, the outer 2 narrow and reflexed, the inner broader and erect; achenes lenticular, 3–5 mm across including the broad, reddish wings. Common. Rocky slopes and ridges, 7,000–8,500 ft. Subalpine F., Alpine Fell-fields. Caribou Mt. (*350*), head of Kidd Cr. (*442*), below ice fields on the north side of Thompson Pk. (*911*), Red Rock Mt. (*975*). August.

Polygonum L.

Fls. many in dense terminal spikes.................... *P. bistortoides*
Fls. 1–several in axillary fascicles or in interrupted spikes.
  Ocrea membranous, turning reddish brown, not lacerate. *P. davisiae*
  Ocrea hyaline, lacerate.
    Fls. drooping on reflexed pedicels.............. *P. douglasii*
    Fls. erect on erect pedicels, or somewhat nodding in age.
      Fls. in axillary fascicles; lvs. ovate......... *P. minimum*
      Fls. in terminal, interrupted, bracteate spikes; lvs. linear to linear-oblanceolate .... *P. spergulariaeforme*

P. bistortoides Pursh. Western Bistort. Perennial from stout, fleshy rootstocks; stems simple, 3–8 dm tall; basal lvs. oblanceolate, 5–30 cm long, tapering to long petioles; cauline lvs. few and much reduced; fls. many in a dense terminal spike; calyx white, ca. 4 mm long; achenes brown, shining, 3–4 mm long. Common. Bogs and wet meadows, 6,500–7,200 ft. Red Fir F. Landers L. (*981*), head of Union Cr. (*994*), Ward L. (*1089*). August.

P. davisiae Brew. ex Gray. Davis' Knotweed. Perennial from a stout, woody, branched taproot; stems several, decumbent to ascending, 1–3 dm long; lvs. sessile, glaucous, glabrous to scabrous, ovate, 1.5–4 cm long; fls. many in small terminal and axillary clusters; calyx greenish white to purplish, 2–3 mm long; achenes light brown, shining, 3–4 mm long. Common. Dry, rocky slopes and ridges, 7,000–8,000 ft. Red Fir F., Subalpine F. Caribou Summit (*186*, *341*), Dorleska Summit (*302*), Willow Cr. Trail (*428*), below ice field on the north side of Thompson Pk. (*906*). July–August.

P. douglasii Greene. Douglas' Knotweed. Slender, glabrous annual; stems simple, 0.5–1 dm tall; lvs. oblanceolate to linear, remote, tapered to base, 13–20 mm long; fls. 1–3 in the axils, drooping; pedicels slender, reflexed, ca. 0.5 mm long; calyx segments green with white margins, ca. 4 mm long; achenes black, shining, ca. 4 mm long. Found at one location. Dry, wooded river bottom, 5,000 ft. Mixed Conifer F. Along the South Fork of the Salmon R. (*1036*). August.

P. minimum Wats. Leafy Dwarf Knotweed. Diffusely branched annual; stems scaberulous, erect, 1–1.5 dm tall; lvs. ovate, 5–10 mm long, not much reduced upwards; fls. many, in fascicles of 2–3 in most lvs.; calyx ca. 2 mm long, the segments green with white margins; achenes black, shining, ca. 2 mm long. Occasional. Gravel washes, 5,000–6,300 ft. Mixed Conifer F. Near head of the South Fork of the Salmon R. (*417*), Boulder Cr. Basin (*1294*). August.

P. spergulariaeforme Meissn. Fall Knotweed. Openly branched annual; stems erect to slightly spreading, scaberulous, 0.5–2 dm tall; lvs. scattered, linear to linear-oblanceolate, 10–20 mm long; fls. many in terminal, elongate, interrupted spikes, erect to somewhat nodding in age on pedicels 1–1.5 mm long; calyx white to rose with dark midveins, 2–3.5 mm long; achenes black, shining, 3–4 mm long. Occasional. Dry meadows, 5,000–5,500 ft. Mixed Conifer F. Big Flat (*249*), head of the South Fork of the Salmon R. (*431*). July–August.

Rumex L.

At least some lvs. hastate; plants dioecious . . . . *R. angiocarpus*
Lvs. not hastate; plants monoecious or perfect . . . . *R. triangulivalvis*

R. angiocarpus Murbeck. Sheep Sorrel. Dioecious perennial from slender, creeping rootstocks; stems 1.5–2.5 dm high; lvs. 3–7 cm long, the lower on long, slender petioles, the upper becoming reduced, at least the lower blades hastate; infl. a dense, narrow panicle, the fls. ± pendulous on slender pedicels 1–2 mm long; inner sepals of female fls. adnate to the ovary which is ca. 1 mm long in fr.; stigmas 3, deeply fringed, deciduous, placed on the angles of the ovary below the summit; sepals of male fls. ca. 1 mm long, often reddish. Occasional. Disturbed places, 5,000–7,000 ft. Mixed Conifer F., Red Fir F. Big Flat (*1177*), head of Kidd Cr. (*1402*). June–July.

R. triangulivalvis (Danser) Rech. f. Willow Dock. Perennial from a stout caudex; stems 2–4 dm tall; lvs. lanceolate to narrowly elliptic, 6–10 cm long, tapering to short petioles, the upper becoming sessile; fls. many in crowded panicles; pedicels slender, ca. 1 mm long; sepals 6 in 2 series, the outer 3 narrow, ca. 1 mm long, the inner 3 broad, ca. 2 mm long and each bearing a corky callosity; achenes brown, ca. 2 mm long. Found at one location. Ephemeral lake bed, 5,000 ft. Mixed Conifer F. Big Flat (*382*). July–August.

Eriogonum compositum

Oxyria digyna

Calyptridium umbellatum

Claytonia nevadensis

PORTULACACEAE

Capsule circumscissile near base; petals 5–10. . . . . *Lewisia*
Capsule opening by 2 or 3 valves; petals mostly 4–5.
Style 1; capsule 2-valved. . . . . *Calyptridium*
Style branches 3; capsule 3-valved.
Plants from thick, fleshy roots or deep-seated corms. . *Claytonia*
Plants from fibrous roots or reproducing by runners or bulblets (except *M. parvifolia*) . . . . *Montia*

Calyptridium Nutt. in T. & G.

C. umbellatum (Torr.) Greene. Pussy Paws. Fig. 59. Perennial from a slender caudex; stems several, scapose, slender, 5–15 cm tall; lvs. mostly basal, spatulate, 1–4 cm long; infl. umbellate-cymose, densely crowded, subcapitate; sepals white to pinkish, 5–8 mm long with broad, scarious margins; petals white to pinkish, 3–6 mm long; capsule ovoid, 3–4 mm long. Common. Dry, sandy to gravelly places, 5,000–8,000 ft. Mixed Conifer F., Red Fir F., Subalpine F. Josephine L. (*77, 295*), Canyon Cr. (*166*), Caribou Mt. (*337*). June–August.

Claytonia L.

Plants from a deep-seated, globose corm. . . . . *C. lanceolata*
Plants from a tangled mass of fleshy rootstocks. . . . . *C. nevadensis*

C. lanceolata Pursh. Western Spring Beauty. Perennial from a deep-seated, globose corm; stems 1–few, 4–15 cm long; basal lvs. 0–2, cauline lvs. several, lanceolate, 3–7 cm long; fls. few-many in terminal racemes; pedicels slender, 1–2 cm long, recurved in fr.; sepals 3–5 mm long; petals white to pink, 6–10 mm long; capsule ca. 4 mm long. Common. Damp woods and meadows, 5,000–7,500 ft. Mixed Conifer F., Red Fir F. Big Flat (*502*), Packers Pk. (*544*), head of Union Cr. (*654*). June–July.

C. nevadensis Wats. Sierra Claytonia. Fig. 60. Perennial from a tangled mass of fleshy rootstocks; stems several, 3–4 cm long; basal lvs. spatulate, the blades 1–3 cm long, 7–20 mm wide, narrowed to slender petioles 2–4 cm long, cauline lvs. ovate, sessile, 6–10 mm long; infl. few-several fld.; pedicels 1–2 cm long; sepals 5–6 mm long; petals 5–8 mm long, white with pink lines; capsule ovoid, 3–3.5 mm long. Rare. Wet, gravelly slopes, 7,800 ft. Alpine Fell-fields. Below ice field on north side of Thompson Pk. (*907*). August. This species has been previously reported for Calif. only from the Sierra Nevada Mts. and Mt. Lassen (Chambers, 1963).

Lewisia Pursh.

Plants without basal lvs. but with 2–5 linear, whorled, cauline lvs. *L. triphylla*
Plants with several–many basal lvs., cauline lvs. reduced to bracts.

Stems scarcely equal to basal lvs.; fls. solitary at the ends of the stems . . . . *L. nevadensis*
Stems greatly exceeding basal lvs.; fls. many in crowded to open panicles.
Lvs. subterete . . . . *L. leana*
Lvs. distinctly flattened.
Petals pink to purplish, 6–10 mm long; basal lvs. mostly less than 1 cm wide; panicle open . . . . *L. columbiana*
Petals white with red stripes or tinged red, 10–14 mm long; basal lvs. broader; panicle short and crowded.
Lvs. entire . . . . *L. cotyledon*
Lvs. strongly toothed on upper margins.
L. cotyledon var. *heckneri*

L. columbiana (Howell) Robinson. Columbia Lewisia. Fig. 61. Perennial from a thick, fleshy caudex; stems few, 10–20 cm tall; basal lvs. many, fleshy, flattened, oblanceolate, 2–6 cm long; cauline lvs. few, reduced to gland-toothed bracts; fls. many in an open panicle with ascending branches; pedicels 3–8 mm long; sepals 2–3 mm long; glandular-dentate; petals 7–10, pink to purplish, 6–10 mm long; capsule ca. equal to sepals. Occasional. Rocky slopes, 6,500–7,700 ft. Red Fir F. Caribou Basin (*190, 744*), Canyon Cr. (*1296*). The nomenclature here follows Abrams (1944), who reports *L. columbiana* from "the mountains of northwestern California." Munz (1959) does not include this species in the Calif. flora.

L. cotyledon (Wats.) Robinson in Gray. Siskiyou Lewisia. Succulent perennial from a heavy caudex; stems few–several, 10–25 cm tall; basal lvs. many, spatulate to oblanceolate, entire, 4–10 cm long; cauline lvs. reduced to glandular-toothed bracts; fls. many in a crowded panicle; pedicels stout, 3–5 mm long; sepals ca. 4 mm long with gland-tipped teeth; petals 5–10, white with reddish tinge or stripes, 10–14 mm long; capsule ovoid, 5–6 mm high. Common. Rocky places, 5,500–8,000 ft. Mixed Conifer F., Red Fir F., Subalpine F. Caribou Basin (*207*). June–July.

var. *heckneri* (Mort.) Munz. Lvs. strongly toothed on the upper margins. Found at one location. Rocky slope, 5,600 ft. Mixed Conifer F. Josephine L. Basin (*76*). June.

L. leana (Porter) Robinson in Gray. Lee's Lewisia. Perennial from a heavy caudex and branched roots; stems few–several, 10–20 cm tall; basal lvs. many, fleshy, terete to somewhat flattened, 2–5 cm long; cauline lvs. few, reduced to gland-toothed bracts; fls. many in an open panicle with ± divergent branches; pedicels slender, 3–12 mm long; sepals ca. 2 mm long, lacerate-dentate, the teeth tipped with dark glands; petals 6–8, mostly pink, 5–7 mm long; capsule ovoid, 4–5 mm high. Common. Rocky places, 5,600–8,400 ft. Mixed Conifer F., Red Fir F., Subalpine F., Alpine Fell-fields. Canyon Cr.

(*168*), Caribou Basin (*208, 735*), Caribou Mt. (*357*), Grizzly L. Basin (*882*). June–August.

L. nevadensis (Gray) Robinson in Gray. Nevada Lewisia. Perennial from a deep-seated fusiform root surmounted by a short caudex; stems few-several, 1–4 cm long, partly underground, bearing a pair of opposite bracts just above the surface; basal lvs. few, linear to ± broadened upward, 3–10 cm long; fls. solitary at the ends of the stems on stout pedicels 3–6 cm long; sepals broad, 6–8 mm long; petals 7–8, white to pinkish, 8–12 mm long; capsule 4–6 mm long. Common. Damp meadows, 5,000–6,600 ft. Mixed Conifer F., Red Fir F. Big Flat (*514, 584*), Dorleska Mine (*614*). June–July.

L. triphylla (Wats.) Three-leaved Lewisia. Perennial from a deep-seated, globose corm; stems slender, 1–few, 4–6 cm long, half underground; basal lvs. none; cauline lvs. 2–5, whorled, linear, 1.5–3 cm long; fls. few-several, subumbellate; pedicels slender, 5–10 mm long; sepals 2–4 mm long; petals white to pinkish, 4–5 mm long; capsule ovoid, 3–4 mm long. Occasional. Damp, gravelly places, 5,500–6,600 ft. Mixed Conifer F., Red Fir F. Josephine L. Basin (*74*), Dorleska Mine (*618*). June–July.

Montia L.

Cauline lvs. alternate . . . . *M. parvifolia*
Cauline lvs. opposite.
  Cauline lvs. 2–several pairs . . . . *M. chamissoi*
  Cauline lvs. 1 pair.
    Cauline lvs. separate . . . . *M. sibirica*
    Cauline lvs. joined at least on one side.
      M. perfoliata var. *depressa*

M. chamissoi (Ledeb.) Dur. & Jacks. Toad-lily. Fig. 62. Perennial producing slender runners; stems ascending to erect, 3–6 cm tall; cauline lvs. opposite in several pairs, oblanceolate, 2–3 cm long; fls. in axillary or subterminal racemes; pedicels slender, 2–10 mm long, recurved in fr.; sepals ca. 2 mm long; petals white, 5–7 mm long; capsule 1–1.5 mm long. Found at one location. Damp meadow, 5,000 ft. Mixed Conifer F. Big Flat (*516*). June.

M. parvifolia (Moc. in DC.) Greene. Small-leaved Montia. Perennial from a thickened caudex; stems slender, erect, 8–20 cm tall; basal lvs. fleshy in dense rosettes, spatulate, 1–2.5 cm long; cauline lvs. alternate, bracteate, reduced upward; fls. 1–10 in terminal racemes; pedicels 5–10 mm long; sepals 2–2.5 mm long; petals white to pinkish with pink or purple veins, 6–8 mm long; capsule ca. 3 mm long. Common. Rocky places, 5,000–6,500 ft. Mixed Conifer F., Red Fir F. Boulder Cr. Basin (*173*), old Caribou Trail (*220*), Josephine L. Basin (*274*), Canyon Cr. (*837*). July–August.

M. perfoliata (Donn) Howell var. *depressa* (Gray) Jeps. Miner's Lettuce. Low reddish annual; stems spreading to erect, 1.5–10 cm long; basal lvs. ovate to deltoid, 5–15 mm broad on long, slender petioles; cauline lvs. united on one side; infl. compact, not much exceeding cauline lvs., fls. 5–10 on short, slender pedicels; sepals 1–2 mm long; petals white, 2–4 mm long; capsule 2–3 mm long. Common. Damp to dry, shaded slopes, 5,000–6,000 ft. Mixed Conifer F. Big Flat (*495, 606*), along South Fork of the Salmon R. (*107*), Union Cr. (*645*). June–July.

M. sibirica (L.) Howell. Siberian Montia. Perennial from a slender rootstock; stems few–several, 10–20 cm tall; basal lvs. few to several, the blades broadly to narrowly elliptic, 2–4 cm long on long, slender petioles; cauline lvs. opposite, in one pair, sessile, distinct, broadly to narrowly ovate, 1.5–3.5 cm long; fls. many in elongate, bracteate racemes; pedicels slender, 1–2.5 cm long, divergent in fr.; sepals 3–6 mm long; petals white with pink lines, 5–8 mm long; capsule shorter than sepals. Common. Damp places, 5,000–6,200 ft. Mixed Conifer F., Red Fir F. Josephine L. Basin (*109*), Big Flat (*144, 490*), along the South Fork of the Salmon R. (*411, 1037*), Kidd Cr. (*692*). June–August.

PRIMULACEAE

Corolla lobes reflexed . . . . *Dodecatheon*
Corolla lobes spreading to erect . . . . *Primula*

Dodecatheon L.

Lvs. spatulate to elliptic . . . . *D. hendersonii*
Lvs. linear to oblanceolate.
  Lvs. linear to linear-oblanceolate, 2–10 mm wide.
   D. alpinum ssp. *majus*
  Lvs. oblanceolate, 8–30 mm wide. . . . . *D. jeffreyi*

D. alpinum (Gray) Greene ssp. *majus* H. J. Thomps. Alpine Shooting Star. Glabrous perennial from a cluster of white, fibrous roots; scapes 1.5–3 dm tall; lvs. linear to linear-oblanceolate, 5–15 cm long, 2–10 mm wide; umbels 3–5 fld., pedicels 1–5 cm long; calyx tube 2–3 mm long, the lanceolate lobes ca. 5 mm long; corolla tube maroon, yellow above, the lanceolate lobes lavender, 10–20 mm long; anthers ca. 6–7 mm long; capsule 6–8 mm long. Found at one location. Wet stream bank, 6,600 ft. Red Fir F. Dorleska Mine (*620*). July.

D. hendersonii Gray. Henderson's Shooting Star. Perennial from a cluster of white fibrous roots producing rice grain bulblets at flowering; scapes mostly glabrous, 1–1.5 dm tall; lvs. spatulate, 2–5 cm long, 8–25 mm wide; umbel 3–7 fld., pedicels 1–4 cm long; calyx tube ca. 2 mm long, the lanceolate lobes 3–5 mm long; corolla tube maroon, yellow above, the lobes lavender, 6–15 mm long; anthers 4–5 mm long; capsule 8–15 mm long. Occasional. Open woods and meadows, 5,000 ft. Mixed Conifer F. Big Flat (*520, 542*). June.

D. jeffreyi Van Houtte. Jeffrey's Shooting Star. Fig. 63. Perennial from a cluster of white, fibrous roots; scapes 1.5–6 dm tall, ± glandular-pubescent above; lvs. oblanceolate, 5–35 cm long, 8–30 mm wide, sparingly glandular-pubescent; umbels 2–10 fld.; pedicels 1–7 cm long; calyx tube 2–5 mm long, the lanceolate lobes 5–10 mm long; corolla tube with a maroon ring below and yellow above, the lobes lavender, 10–20 mm long; anthers 7–9 mm long; capsule 7–12 mm long. Common. Damp to wet meadows and bogs, 5,000–8,400 ft. Mixed Conifer F., Red Fir F., Subalpine F., Alpine Fell-fields. Canyon Cr. (*152, 800, 821, 852*), Caribou Basin (*202, 727*), below ice fields on north side of Thompson Pk. (*903*), head of Union Cr. (*650*), Josephine L. Basin (*72*). June–August.

Primula L.

P. suffrutescens Gray. Sierra Primrose. Perennial from slender rhizomes; scapes 5–10 cm long, glandular-puberulent above; lvs. crowded, 2–3 cm long, rounded and dentate at the apex, becoming gradually narrowed to broad petioles; umbels 4–7 fld., pedicels 1–1.5 cm long; calyx tube ca. 3 mm long, the lanceolate lobes ca. 4 mm long; corolla magenta with yellow throat, the tube 8–10 mm long, the spreading to erect lobes ca. as long; capsule ovoid, scarcely as long as calyx. Common. Damp slopes, 7,200–8,400 ft. Subalpine F., Alpine Fell-fields. Caribou Mt. (*354*), Caribou Basin (*399, 1045*), below ice fields on north side of Thompson Pk. (*922*). August.

PYROLACEAE

Plants fleshy stemmed from thickened bases; without green lvs. *Pterospora*

Plants from trailing rootstocks; mostly with green lvs.

- Fls. in racemes, several to many .... *Pyrola*
- Fls. in corymbs, 1–7 .... *Chimaphila*

Chimaphila Pursh

Lvs. ovate to elliptic; peduncle with 1–3 white fls......... *C. menziesii*

Lvs. oblanceolate; peduncle with 3–7 pink fls. *C. umbellata* var. *occidentalis*

C. menziesii (R. Br. ex D. Don) Spreng. Little Prince's Pine. Low, perennial herb from slender, trailing underground rootstocks; lvs. ovate to narrowly elliptic, 1–5 cm long, serrulate, dark green above, paler beneath; peduncles 3–5 cm long, 2–3 fld.; sepals white, rounded, erose, veiny, ca. 5 mm long; petals white to pinkish in age, round, ca. 6 mm long; filaments with a dilated, pubescent base; capsule globose, 5–6 mm in diam. Occasional. Shaded woods, 5,000 ft. Mixed Conifer F. Along the South Fork of the Salmon R. (*270*). July–August.

61

Lewisia columbiana

62

Montia chamissoi

63

Dodecatheon jeffreyi

64

Chimaphila umbellata var. occidentalis

C. umbellata (L.) Barton var. *occidentalis* (Rydb.) Blake. Western Prince's Pine. Fig. 64. Low perennial herb from thickened, trailing, underground rootstocks; lvs. oblanceolate, 1–6 cm long, ± whorled, sharply serrate, dark green above, paler beneath; peduncles 4–6 cm tall, 3–7 fld.; sepals purplish, ovate, fimbriate, 3–4 mm long; petals pink to purplish, ciliolate, 5–6 mm long; filaments with a dilated, pubescent base; capsule globose, 6–7 mm in diam. Common. Dry, wooded slopes, 5,000–7,000 ft. Mixed Conifer F., Red Fir F. Along the South Fork of the Salmon R. (*271*), Boulder Cr. Basin (*481*), Big Flat (*709*), Canyon Cr. near "L" Lake (*825*). July–August.

Pterospora Nutt.

P. andromedea Nutt. Pinedrops. Stout, simple, clammy-pubescent root-parasite, lacking chlorophyll; stems 3–10 dm tall; lvs. linear, crowded toward base of stem, 2–4 cm long; infl. an elongate, many-fld. raceme; the pedicels reflexed in fr.; sepals reddish, glandular-ciliate, 4–5 mm long; petals white to reddish, ca. 7 mm long; anthers bearing long, slender awns; capsule 8–12 mm in diam. Common. Shaded woods, 5,000–5,600 ft. Mixed Conifer F. Along South Fork of the Salmon R. (*285*), Kidd Cr. Trail (*445*), Grizzly Cr. (*930*). July–August.

Pyrola L.

Style deflexed; infl. not 1-sided.
  Lvs. well developed.
    Petals pink to purple; lvs. orbicular. . . . *P. asarifolia* var. *purpurea*
    Petals mostly white to greenish; lvs. ovate to broadly elliptic. . . . *P. picta*
  Lvs. reduced to scales 1–2 cm long. . . . . *P. picta* forma *aphylla*
Style straight; infl. 1-sided. . . . . *P. secunda*

P. asarifolia Michx. var. *purpurea* (Bunge) Fern. Bog Wintergreen. Scapose perennial from slender rootstocks; scapes 3–4 dm tall, ± twisted, bearing 2–3 lanceolate bracts; lvs. basal, the blades orbicular on petioles 2.5–7 cm long; racemes several fld.; pedicels recurved in age; sepals pinkish, ca. 2 mm long, lanceolate; petals pink to purple, ca. 5 mm long; styles deflexed; capsule ca. 5 mm broad. Found at one location. Wet, shaded place, 5,700 ft. Red Fir F. Canyon Cr. above the upper lake (*794*). July–August.

P. picta Sm. White-veined Wintergreen. Fig. 65. Scapose perennial from slender, creeping rootstocks; scapes 1–2.5 dm tall, bearing 1–3 bracts; lvs. basal, the blades ovate to broadly elliptic, white veined or mottled, serrulate to entire, 2–4.5 cm long on petioles 1–2 cm long; racemes several–many fld.; pedicels recurved in age; sepals green with hyaline margins, ca. 2 mm long; petals greenish to white, 6–7 mm long; styles deflexed; capsule 5–6 mm broad. Common. Shaded woods, 5,000–7,500 ft. Mixed Conifer F., Red Fir F. South Fork Trail (*320*), Caribou Mt. (*326, 715*), Canyon Cr. (*828*), Sawtooth Ridge (*1073*). July–August.

forma *aphylla* (Sm.) Camp. Lvs. of flowering stalks scalelike, of sterile stalks sometimes green, 1–2 cm long; sepals red-purple, ca. 2 mm long; petals white, green, or pink. Found at one location. Shaded woods, 5,000 ft. Mixed Conifer F. Along the South Fork of the Salmon R. (*268*). July–August.

P. secunda L. One-sided Wintergreen. Scapose perennial from slender, trailing rootstocks; scapes 0.7–1 dm tall; lvs. basal, the blades ovate, 1–4 cm long on petioles 1–2 cm long; racemes several fld., one-sided; pedicels recurved in age; sepals green to somewhat purplish, 1–1.5 mm long; petals white to greenish, 4–5 mm long; styles straight; capsule 4–5 mm broad. Common. Shaded woods, 5,000–7,500 ft. Mixed Conifer F., Red Fir F., Subalpine F. Along South Fork of the Salmon R. (*269*), Caribou Basin (*394*), Caribou Mt. (*714*), Canyon Cr. (*829*), Sawtooth Ridge (*1072*). July–August.

RANUNCULACEAE

Fls. irregular.
- Upper sepal projected backward into an elongate spur. . *Delphinium*
- Upper sepal hooded . . . . *Aconitum*

Fls. regular.
- Petals produced backward into long spurs . . . . *Aquilegia*
- Petals not spurred.
  - Pistils few to several ovuled, 6–15 mm long in fr.
    - Sepals 10–15 mm long; lvs. simple . . . . *Caltha*
    - Sepals 2–3 mm long; lvs. compound. . . . . *Actaea*
  - Pistils 1–ovuled, less than 4 mm long, excluding style.
    - Sepals spurred; receptacle several times longer than broad in fr. . . . . *Myosurus*
    - Sepals not spurred; receptacle not greatly elongate.
      - Fls. white to bluish . . . . *Anemone*
      - Fls. yellow to pale yellow . . . . *Ranunculus*

Aconitum L.

A. columbianum Nutt. Columbia Monkshood. Perennial from a tuber 1–3 cm long; stems stout, erect, 5–20 dm tall, glabrous below, becoming hirsute above; lf. blades 5–10 cm broad, deeply palmately divided, the primary divisions again divided into narrow segments; petioles slender, the lower much longer than the blades, the upper becoming shorter than the blades; infl. open, several fld.; fls. white to blue; sepals 5, the upper produced into a narrow, beaked hood 15–20 mm high, the lateral pair ± rounded, 10–15 mm long, the lower pair linear, 7–10 mm long; petals 2, pale, enclosed by the hood, clawed, ca. 15 mm long; stamens many, filaments expanded at the base; pistils 3, 12–18 mm long in fr. Common. Damp meadows and shaded places, 5,000–6,200 ft. Mixed Conifer F., Red Fir F. Josephine L. (*291, 292*), Kidd Cr. (*437, 438*), Canyon Cr. (*785, 854*), Grizzly Cr. (*870*). August.

Actaea L.

A. rubra (Ait.) Willd. ssp. *arguta* (Nutt.) Hult. Western Red Baneberry. Glabrous perennial from a branching rootstock; stems 3–6 dm high; lvs. large, 2–3 in number, ternately divided; lfts, thin, ovate, with coarsely toothed margins, dark green, 3–8 cm long; fls. in a terminal raceme; sepals and petals white, 2–3 mm long, early deciduous; stamens numerous; berries red at maturity 7–10 mm long. Occasional. Rocky slopes and meadows, 5,500–6,500 ft. Mixed Conifer F., Red Fir F. Boulder Cr. Basin (*1292a*), Canyon Cr. (*Alexander & Kellogg 5420*). June–July.

Anemone L.

Lvs. entire to broadly lobed; plants from slender, creeping rootstocks. *A. deltoidea*

Lvs. greatly dissected into linear segments; plants from stout, woody rootstocks.

Sepals 10–16 mm long; styles 1.5–3 mm long in fr... *A. drummondii*

Sepals 25–40 mm long; styles 15–30 mm long in fr... *A. occidentalis*

A. deltoidea Hook. Columbia Wind-flower. Fig. 66. Perennial from slender, creeping rootstocks; stems slender, 7–15 cm tall; basal lvs. 3-foliolate, the lfts. ovate, 3–4 cm long, crenate, petioles slender, ca. as long as stem; cauline lvs. simple, 3 in a whorl, ovate, crenate to 3 lobed, 3–5 cm long, the petioles 2–5 mm long; peduncles 1.5–9 cm long, bearing a solitary fl.; sepals 5, white, 10–18 mm long; petals absent; stamens many; pistils many, hairy; achenes ca. 3 mm long, hirsute, with a beak ca. 1 mm long. Common. Shaded woods, 5,000–6,000 ft. Mixed Conifer F., Red Fir F. Big Flat (*89*), Josephine L. Basin (*112*), Yellow Rose Mine Trail (*135*), Kidd Cr. (*433*). June–August.

A. drummondii Wats. Drummond's Anemone. Perennial from a stout root-crown surmounting a tough taproot; stems erect, 5–20 cm tall, glabrous to sparingly villous; basal lvs. on long, slender petioles, the blades greatly dissected into linear segments, sparingly villous; involucral lvs. 3, on short, broad petioles, the blades dissected; fls. solitary on a long, slender peduncle, 5–20 cm long, 0.5–1 mm thick; sepals 5–8, white, 10–16 mm long; petals absent; stamens many; pistils many, densely woolly; styles 1.5–3 mm long in fr. Common. Dry, rocky slopes and meadows, 6,500–8,000 ft. Red Fir F., Subalpine F. Dorleska Summit (*120, 317*), Yellow Rose Mine (*573*), Union L. (*641*), head of Union Cr. (*1004*). June–July.

A. occidentalis Wats. Western Pasque-flower. Perennial from a stout, branched, woody caudex; stems 15–25 cm tall, nearly glabrous to villous; basal lvs. on long, slender petioles, the blades greatly dissected into linear segments, involucral lvs. 3, on shorter, slender petioles, the blades dissected; fls. solitary on a stout peduncle 10–30 cm long, 2–2.5 mm thick; sepals 5–8, white to bluish, 25–40 mm long; petals none; stamens many; pistils many, densely woolly; styles densely plumose, 15–30 mm long in fr. forming soft, globose heads. Common. Dry, rocky slopes, 6,500–8,000 ft. Red Fir F., Subalpine F. Caribou Basin (*754*), Canyon Cr. (*809*), below ice fields on the north side of Thompson Pk. (*925*). July–August.

Aquilegia L.

A. formosa Fisch. in DC. Northwest Crimson Columbine. Perennial from a thick, heavy caudex; stems 3–7 dm tall, branched, glabrous and glaucous below, becoming pubescent above; basal lvs. biternate, on long, slender petioles, the lfts. 1–4 cm long, crenate to deeply lobed; cauline lvs. becoming reduced upwards; fls. nodding, borne at the ends of the branches; sepals reddish, 10–15 mm long; petals produced forward into a yellow lamina and produced backward into a tubular, reddish spur 10–20 mm long; stamens many, the innermost represented by a pale sheath of flattened staminodia; pistils 5; follicles glandular-pubescent, 10–20 mm long, the slender styles an additional 5–15 mm long. Common. Stream banks and damp, shaded places, 5,500–7,000 ft. Red Fir F. Josephine L. Basin (*273*), Caribou Basin (*392, 722*), Canyon Cr. (*799*), Landers L. (*1010*). July–August.

Caltha L.

C. howellii (Huth) Greene. White Marsh-marigold. Perennial from short rootstocks with fibrous roots; lvs. basal, the blades round-reniform, 3–10 cm wide, the petioles 7–25 cm long; fls. solitary on a naked scape 10–30 cm long; sepals 5–9, greenish to white, 10–15 mm long; petals absent; stamens many; pistils several; follicles 10–15 mm long. Common. Bogs, 6,000–7,400 ft. Red Fir F. Below Dorleska Mine (*624*), head of Union Cr. (*649*). July–August.

Delphinium L.

Stems simple, 10–20 dm tall, not attenuate at base . . . . *D. glaucum*
Stems branched, 1.5–5.5 dm tall, attenuate at base . . . . . *D. nuttallianum*

D. glaucum Wats. Glaucous Rocky Mountain Larkspur. Perennial from a stout woody caudex; stems erect, simple, 10–20 dm tall; glabrous and glaucous; basal lvs. 10–15 cm broad, pubescent, deeply palmately lobed, the primary divisions ± incised, petioles 10–15 cm long; fls. many in a crowded, elongate raceme; sepals blue, pubescent, the spur ca. 10 mm long, the lateral and lower sepals 7–8 mm long; upper petals whitish, glabrous, lower petals blue, densely bearded; follicles glabrous to puberulent, 10–15 mm long. Found at one location. Stream bank, 5,500 ft. Mixed Conifer F. Near the head of the South Fork of the Salmon R. (*404*). August.

D. nuttallianum Pritz. ex Walp. Dwarf Larkspur. Perennial from a tuberous root; stems 1.5–5.5 dm tall, glabrous, branched; lvs. 3–7 cm broad, glabrous to somewhat pubescent, palmately lobed, the primary divisions again divided, petioles 4–12 cm long; infl. open, few–several fld.; sepals blue, pubescent, the spur 10–15 mm long, the lateral and lower sepals 10–12 mm long; upper petals whitish, the lower ones blue, bearded; follicles glabrous to densely hirsute, 10–20 mm long. Common. Open to shaded rocky meadows and slopes, 5,000–7,000 ft. Mixed Conifer F., Red Fir F. Big Flat (*150*, *581*), Canyon Cr. (*157*, *805*), Caribou Basin (*201*, *743*), Dorleska Summit (*577*), Bullards Basin (*629*), Union Cr. (*646*), along the South Fork of the Salmon R. (*88*), Dorleska Mine (*617*). June–August. Plants with glabrous follicles resemble *D. depauperatum* but because of their consistently larger and hairy sepals and usually taller stems they have been included here.

Myosurus L.

M. minimus L. ssp. *montanus* Camp. Common Mouse-tail. Low annuals 2–7 cm high; lvs. basal, linear-filiform to linear-oblanceolate, 2–4 cm long; fls. solitary on scapes exceeding the lvs.; sepals greenish, ca. 2 mm long with a basal, hyaline spur ca. 1 mm long; petals when present greenish-yellow; stamens 5–10; receptacle 2–3 mm wide and 10–20 mm long in fr.; beak of achenes 0.5–1 mm long, erect to ± divergent. Occasional. Wet meadows, 5,000 ft. Mixed Conifer F. Big Flat (*1192*). June.

Ranunculus L.

Plants glabrous.
- Basal lvs. 3-parted . . . . *R. eschscholtzii*
- Basal lvs. entire . . . . *R. alismaefolius* var. *hartwegii*

Plants hirsute, sometimes thinly so.
- Petals 7–12 mm long . . . . *R. orthorhynchus*
- Petals ca. 3 mm long . . . . *R. uncinatus*

R. alismaefolius Geyer ex Benth. var. *hartwegii* (Greene) Jeps. Water Plantain Buttercup. Glabrous perennial from several thickened, fibrous roots; stems ascending to erect, 1–2 dm long; lvs. mostly basal, the blades lanceolate, 3–6 cm long, 5–10 mm wide, tapering to slender petioles; fls. borne on rather stout pedicels 2–10 cm long; sepals greenish, ca. 4 mm long; petals 5–6, bright yellow, 6–8 mm long; achenes 2–2.5 mm long, glabrous, with a beak ca. 0.5 mm long. Occasional. Open, damp, rocky meadows, 6,500 ft. Red Fir F. Dorleska Mine (*613*), head of Union Cr. (*652*). July.

R. eschscholtzii Schlecht. Eschscholtz's Buttercup. Glabrous perennial from a caudex bearing slender roots; stems decumbent to erect, 1–1.5 dm long; basal lvs. 2–3 cm broad on slender petioles, deeply 3-lobed, the primary lobes again ± divided; cauline lvs. sessile, deeply lobed; fls. solitary on stout pedicels 2–10 cm long; sepals greenish and tinged lavender, 4–5 mm long, ± pubescent; petals 5, yellow, 6–8 mm long; achenes ca. 1.5 mm long with a slender beak ca. 0.5 mm long, borne in an ovoid head 5–12 mm long. Occasional. Wet, rocky slopes, 5,000–8,000 ft. Red Fir F., Subalpine F., Alpine Fell-fields. Caribou Basin (*758*, *1051*), below ice fields on the north side of Thompson Pk. (*915*). July–August.

Pyrola picta

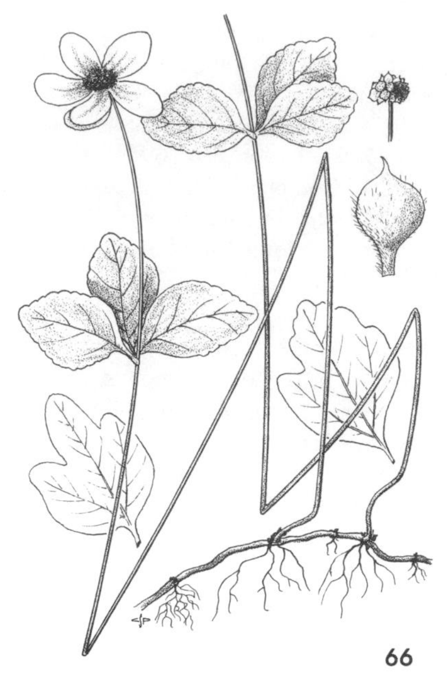

Anemone deltoidea

Ranunculus orthorhynchus

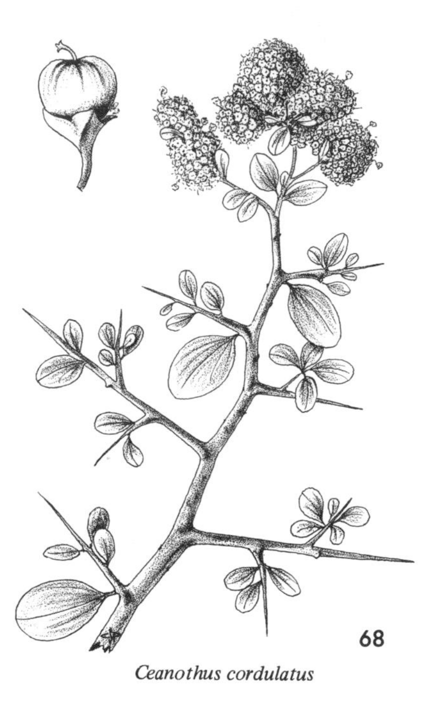

Ceanothus cordulatus

R. orthorhynchus Hook. Straight-beaked Buttercup. Fig. 67. Perennial from a mass of somewhat fleshy roots; stems spreading to erect, 1–3.5 dm tall, hirsute; basal lvs. pinnate with 5–7 lfts., these ± divided and hirsute; fls. solitary on hirsute pedicels 2–5 cm long; sepals pilose, 5–6 mm long, reflexed; petals 5, yellow to whitish, 7–12 mm long; achenes 2–4 mm long with a beak 4–5 mm long. Common. Dry, open meadows, 5,000 ft. Mixed Conifer F. Big Flat (*145, 522*), along the South Fork of the Salmon R. (*498, 590*), head of the South Fork of the Salmon R. (*106*). June–July.

R. uncinatus D. Don in G. Don. Bongard's Buttercup. Perennial from a cluster of fibrous roots; stems 2–4 dm tall, thinly hirsute, with slender branches above; basal lvs. cordate-reniform in outline, 3-parted, 2–9 cm broad; cauline lvs. sparingly hirsute, becoming reduced above; fls. borne on slender pedicels 2–4 cm long; sepals greenish, ca. 2 mm long; petals 5, pale yellow, ca. 3 mm long; achenes ca. 3 mm long with a hooked beak ca. 1 mm long. Occasional. Open meadows, 5,000–7,000 ft. Mixed Conifer F., Subalpine F. Head of Kidd Cr. (*1109*), Big Flat (*1180*). June–August.

RHAMNACEAE

Mature fr. a dry capsule . . . . *Ceanothus*
Mature fr. fleshy, berrylike . . . . *Rhamnus*

Ceanothus L.

Ultimate branches stiff, spinose . . . . *C. cordulatus*
Ultimate branches flexible, not spinose.
Prostrate shrubs with blue fls.; lvs. sharply toothed toward apex. *C. prostratus*
Erect, spreading shrubs with white fls.; lvs. glandular-serrate. *C. velutinus*

C. cordulatus Kell. Mountain Whitethorn. Fig. 68. Low, spreading shrub 5–15 dm tall; bark smooth, light gray; ultimate branches stiff and spinose; lvs. alternate, ovate, elliptic or obovate, 1–2 cm long with 3 prominent veins, green above, grayer beneath, petioles slender, 1–4 mm long; fls. white in dense clusters 1.5–3 cm long; capsule triangular, mostly lobed, 4–5 mm broad. Common. Dry slopes and flats, 5,000–6,000 ft. Mixed Conifer F. Big Flat (*582, 603*). June–July.

C. prostratus Benth. Mahala-mats. Prostrate, mat-forming shrub; bark rough, brownish gray; branches not spinose; lvs. opposite, cuneate-oblanceolate to obovate, 1–2 cm long, becoming sharply toothed toward the apex, shiny green above, canescent below, with many lateral veins; fls. blue in dense clusters 1–2 cm long; capsule subglobose, 5–9 mm broad. Occasional. Wooded ridges and slopes, 6,500–7,000 ft. Red Fir F. Ridge between Landers L. and the head of Union Cr. (*659*). July.

C. velutinus Dougl. ex Hook. Tobacco-brush. Spreading, rounded shrubs, 1–2 m tall; bark greenish to brown; ultimate branches flexible, not spinose; lvs. alternate, elliptic to ovate, 2.5–6 cm long, glandular-serrate, glossy green above, paler and finely canescent beneath, with 3 prominent veins; fls. white in dense clusters 3–8 cm long; capsule subglobose to triangular, 3–4 mm wide. Common. Dry, open slopes, 5,000–8,000 ft. Montane Chaparral, Mixed Conifer F., Red Fir F., Subalpine F. Big Flat (*148*, *525*), Caribou Mt. (*718*). June–July.

Rhamnus L.

R. purshiana DC. Cascara Sagrada. Deciduous shrub, 1–2 m high with smooth, gray bark; lvs. thin, elliptic to obovate, dark green, prominently veined, serrulate, 3–7 cm long on petioles 8–12 mm long; fls. ca. 4 mm long, petals greenish; berry black, 8–10 mm in diam., 2–3 seeded. Occasional. Rocky slopes, 5,000–6,000 ft. Mixed Conifer F., Red Fir F. Boulder Cr. Basin (*1280*, *1419*), dry rocks north of Mirror L. (*J. O. Sawyer 2373*). July.

ROSACEAE

Ovary inferior; fr. a pome.
- Lvs. simple . . . . *Amelanchier*
- Lvs. pinnate . . . . *Sorbus*

Ovary or ovaries superior; fr. not a pome.
- Plants herbaceous.
  - Petals yellow.
    - Stamens 5 . . . . *Sibbaldia*
    - Stamens 20–25 . . . . *Potentilla*
  - Petals white.
    - Stamens 10, filaments dilated at base . . . . *Horkelia*
    - Stamens 20–30, filaments not dilated.
      - Fr. of 5 follicles; lvs. dissected into linear divisions. *Luetkea*
      - Fr. of many achenes imbedded in an enlarged, fleshy receptacle; lvs. not dissected into linear divisions. *Fragaria*
- Plants shrubs, trees, or vines armed with prickles.
  - Petals absent; styles 4–7 cm long, silky-plumose . . *Cercocarpus*
  - Petals present; styles much shorter, not as above.
    - Petals yellow.
      - Pistil 1; lvs. simple . . . . *Purshia*
      - Pistils many; lvs. pinnate . . . . *Potentilla*
    - Petals not yellow.
      - Petals white.
        - Pistil 1 . . . . *Prunus*
        - Pistils 5-many.
          - Pistils 5; fr. of dry achenes; infl. a dense panicle . . . . *Holodiscus*

Pistils many; fr. an aggregate of ± fleshy drupelets; infl. not a dense panicle. . *Rubus*

Petals rose to dark purple.

Stems armed with prickles . . . . *Rosa*

Stems not armed . . . . *Spiraea*

Amelanchier Medic.

Lf. blades puberulent to tomentulose, especially below. . . . . . *A. pallida*

Lf. blades mostly glabrous . . . . *A. pumila*

A. pallida Greene. Pallid Service-berry. Deciduous shrub, 1–2 m tall, with erect to spreading branches; bark glabrous, red-brown to gray; lvs. elliptic, oblong, or ovate to roundish, 1–3 cm long, 8–22 mm wide, puberulent to tomentulose, especially on lower surface, usually paler on lower surface, toothed, especially above the middle; racemes pubescent, 4–6 fld.; sepals ca. 2 mm long, ± tomentose without, densely so within; petals white, erect, 8–11 mm long; stamens ca. 15; styles 3–4; fruit a pome, 4–6 mm in diam., purplish black. Common. Open slopes and meadows, 5,000–7,000 ft. Mixed Conifer F., Red Fir F. Big Flat (*149, 493*). June–July.

A. cf. *pumila* Nutt. Smooth Service-berry. Low, decumbent, deciduous shrub to 2 dm high; bark reddish brown to gray; lvs. elliptic, oblong, or ovate, ca. 1 cm long, 5–10 mm wide, serrate to ca. the middle, the blades glabrous except for a few hairs along the midrib, the petioles and stipules sparingly long-pilose; racemes 4–6 fld., the bracts ± sparingly pilose; sepals 1.5–2.5 mm long, glabrous; petals white, 8–10 mm long; stamens 9–15; styles mostly 5, ± united at base; fr. a pome, 8–9 mm in diam., dark purple. Found at one location. Open, rocky meadow, 6,400 ft. Red Fir F. Head of Bullards Basin (*633b*). July. *A. pumila* is described as being completely glabrous and is reported for Calif. only from the Sierra Nevada (Munz, 1959).

Cercocarpus HBK.

C. ledifolius Nutt. Curl-leaved Mountain-mahogany. Fig. 69. Tree with spreading, gnarled branches, 2–3 m tall; bark brownish gray, deeply furrowed; lvs. resinous, lance-elliptic to lanceolate, 1–2.5 cm long, tomentulose to glabrous above, tomentulose below, revolute; fls. mostly 2–4 in a fascicle; sepals ca. 2 mm long, woolly; petals absent; stamens 18–25; pistil 1; fr. a woolly, cylindric, fusiform achene, 5–8 mm long with a silky-plumose style 4–7 cm long. Occasional. Dry, rocky slopes and ridges, 6,500–7,000 ft. Red Fir F. Caribou Mt. (*762*), Packers Pk. (*938*). June–August.

Fragaria L.

F. vesca L. ssp. *californica* Staudt. California Strawberry. Scapose perennial; lvs. pinnately compound, lfts. 3, silky-pubescent beneath, glabrate above,

1.5–3 cm long, coarsely serrate; petioles 2.5–5 cm long, pilose; peduncles several, usually few-fld.; sepals 4–5 mm long, villous, alternating with smaller bractlets; petals white, 5–8 mm long; stamens 25–30; pistils many; achenes borne on the surface of the enlarged, fleshy receptacle. Occasional. Dry slopes and woods, 5,000 ft. Mixed Conifer F. Along the South Fork of the Salmon R. (*103, 110*), Big Flat (*487*). June.

Holodiscus Maxim.

H. microphyllus Rydb. Ocean Spray. Bushy shrub, 3–10 dm tall; bark brownish to gray; lvs. obovate, often cuneate at base, 1–3 cm long, toothed at top to ca. the middle, ± pubescent and green above, paler beneath and glandular-tomentose; infl. compact to diffuse, many fld.; sepals ca. 2 mm long, pubescent; petals white, ca. 2 mm long; stamens ca. 20; pistils 5, the carpels silky-villous; achenes laterally flattened. Common. Dry slopes and ridges, 5,600–7,800 ft. Mixed Conifer F., Red Fir F. Dorleska Summit (*305*), Willow Cr. Trail (*429*), Canyon Cr. (*777*), Packers Pk. (*931*), Sawtooth Ridge (*1044*). August.

Horkelia Cham. & Schlecht.

H. fusca Lindl. ssp. *parviflora* (Nutt.) Keck. Dusky Horkelia. Perennial from a short, stout caudex; stems few, slender, 1–3 dm tall, glandular-pubescent; basal lvs. pinnate with ca. 5 pairs of cuneate to obovate lfts. that are toothed to narrowly lobed at the apex and thinly glandular-villous, cauline lvs. few and with narrower divisions; cymes several fld., compact; sepals 2–3 mm long, thinly villous, alternating with shorter, narrower bractlets; petals white, 4–5 mm long; stamens 10, the filaments dilated at the base; pistils 10–15; achenes brown, ca. 1 mm long. Common. Dry flats and meadows, 5,000 ft. Mixed Conifer F. Big Flat (*69, 250*). July–August.

Luetkea Bong.

L. pectinata (Pursh) Kuntze. Luetkea. Fig. 70. Decumbent perennial, somewhat woody toward the base; flowering stems 5–15 cm tall; lvs. ca. 1 cm long with ultimate, narrow divisions; racemes narrow, 1–5 cm long, ± pubescent; sepals 2 mm long; petals white, ca. 2 mm long; stamens ca. 20, the filaments united at the base; pistils 5; follicles ca. 2.5 mm long. Common. Open, rocky, damp slopes, 6,500–8,200 ft. Red Fir F., Subalpine F., Alpine Fell-fields. Caribou Mt. (*343*), Caribou Basin (*389, 1049*), Canyon Cr. (*815*). August.

Potentilla L.

Plants bushy shrubs . . . . *P. fruticosa*
Plants herbaceous.
    Lvs. pinnate, with 5 or more lfts.
        Lvs. whitish with a dense tomentum; styles filiform. . *P. breweri*

Lvs. green, ± pilose; styles fusiform.
P. glandulosa ssp. *nevadensis*

Lvs. palmate or if pinnate, with 3 lfts.

Lvs. palmate with 4–5 oblanceolate, hirsute lfts.
P. gracilis ssp. *nuttallii*

Lvs. pinnate with 3 flabelliform, glabrous lfts. . . *P. flabellifolia*

P. breweri Wats. Brewer's Cinquefoil. Perennial from a stout, woody caudex; stems 8–12 cm tall, tomentose; lvs. 3–5 cm long, pinnate, the lfts. in 4–6 pairs, crowded, fan-shaped, palmately cleft, densely white-tomentose, cauline lvs. few, 1–5 foliolate; cyme few fld., loose; pedicels 1–2 cm long; sepals 5–6 mm long, tomentose; petals yellow, 6–8 mm long; stamens ca. 20; pistils many, styles filiform, ± terminal; achenes 0.8 mm long, brownish. Found at one location. Rocky slope, 7,400 ft. Red Fir F. Head of Kidd Cr. (*1096*). August.

P. flabellifolia Hook. ex T. & G. Mount Rainier Cinquefoil. Perennial from a scaly, branched rootstock; stems 10–45 cm tall, glabrous below, becoming thinly tomentose above; lvs. mostly basal, 3 foliolate, the lfts. flabelliform, coarsely toothed, glabrous, on long, slender petioles; cyme few fld., compact to spreading; sepals ca. 5 mm long, thinly tomentose; petals yellow, ca. 7 mm long; stamens ca. 20; pistils many, styles filiform, ± terminal; achenes 1.2 mm long, red-brown. Common. Damp, rocky slopes, 6,200–7,800 ft. Red Fir F., Subalpine F., Alpine Fell-fields. Caribou Mt. (*335*, *359*), Boulder Cr. Basin (*480*), Caribou Basin (*390*, *732*), below ice field on north side of Thompson Pk. (*819*). July–August.

P. fruticosa L. Shrubby Cinquefoil. Low, bushy shrub, 20–50 cm tall; bark brownish, ± exfoliating in age; lvs. pinnate, the lfts. 3–7, elliptic to narrowly oblanceolate, silky-pubescent; fls. 1–few in small cymes; sepals ca. 6 mm long, pilose; petals yellow, rounded, 6–10 mm long; stamens ca. 25; pistils many, styles somewhat thickened, lateral; achenes ca. 1 mm long, densely silky-pilose. Common. Open slopes and meadows, 6,500–7,000 ft. Red Fir F. Dorleska Mine (*313*), head of Kidd Cr. (*441*), Grizzly L. Basin (*897*), Landers L. (*980*). August.

P. glandulosa Lindl. ssp. *nevadensis* (Wats.) Keck. Sticky Cinquefoil. Perennial from a woody caudex; stems 10–40 cm tall, densely hirsute throughout to sparsely pubescent below; lvs. pinnate, lfts. 7–9, coarsely toothed, obovate, cuneate at base, pilose, cauline lvs. few, reduced; cymes several fld.; sepals 3–8 mm long, pubescent; petals yellow, 5–10 mm long; stamens ca. 25; pistils many, styles fusiform, basal; achenes brownish, ca. 1 mm long. Common. Open woods, meadows, and slopes, 5,000–7,000 ft. Mixed Conifer F., Red Fir F. Big Flat (*92*, *521*), Canyon Cr. (*160*, *790*, *853*), Caribou Basin (*198*), Josephine L. (*598*), Landers L. (*984*). June–August.

P. gracilis Dougl. ex Hook. ssp. *nuttallii* (Lehm.) Keck. Slender Cinquefoil. Perennial from a stout, woody caudex; stems 3–5 dm tall, hirsute; lvs. palmate, lfts. oblanceolate, 3–5 cm long, coarsely serrate, hirsute, sometimes densely so and appearing grayish; cyme several fld.; sepals 5–7 mm long; petals yellow, 5–10 mm long; stamens ca. 20; pistils many, styles filiform, terminal; achenes greenish brown, 1–1.3 mm long. Common. Dry, rocky meadows and slopes, 5,000–6,700 ft. Mixed Conifer F., Red Fir F. Big Flat (*260*), Dorleska Mine (*312*), ridge west of Bullards Basin (*966*), head of South Fork of Salmon R. (*423*), head of Union Cr. (*1006*). August.

Prunus L.

Fls. 3–10 in ± flat-topped corymbs . . . . *P. emarginata*
Fls. 12–many in elongate racemes . . . . *P. virginiana* var. *demissa*

P. emarginata (Dougl.) Walp. Bitter Cherry. Deciduous shrub, 1–3 m tall; bark reddish on young twigs, becoming gray on older branches; lvs. borne in fascicles, lanceolate, elliptic, or obovate, serrulate, 1–4 cm long, 0.5–1.5 cm wide, mostly glabrous; corymbs mostly 4–5 fld.; sepals ca. 2.5 mm long, soon becoming reflexed; petals white, 4.5–6 mm long; stamens 30–40; pistil 1; drupe red, 6–8 mm in diam. Common. Open meadows and ridges, 5,000–8,000 ft. Mixed Conifer F., Red Fir F., Subalpine F. Yellow Rose Mine (*123*), Big Flat (*532*), Bullards Basin (*632*), Caribou Basin (*716*). June–August.

P. virginiana L. var. *demissa* (Nutt.) Sarg. Western Choke Cherry. Deciduous shrubs or small trees, 1–5 m tall; bark gray-brown, the young twigs pubescent; lvs. borne separately, oblong-ovate to elliptic, serrulate, 2.5–6 cm long, 1.5–3.5 cm broad, ± pubescent beneath, glabrous above; racemes elongate, many fld.; sepals ca. 1 mm long, erect to spreading; petals white, ca. 4 mm long; stamens many; pistils 1; drupe dark red, 5–6 mm thick. Occasional. Dry slopes and woods, 5,000–6,000 ft. Mixed Conifer F. Big Flat (*583*). June–July.

Purshia DC. ex Poir.

P. tridentata (Pursh) DC. Northern Antelope Bush. Low shrub 3–10 dm tall; bark gray, young twigs tomentose; lvs. cuneate, ca. 1 cm long, 3-cleft with revolute margins, green and thinly tomentose above, whitish with a dense tomentum below; fl. tube funnelform, ca. 3 mm long, white-tomentose and stipitate-glandular; sepals tomentose, 1–2 mm long; petals yellow, 4–5 mm long; stamens ca. 25; pistil 1, densely pilose; achene fusiform, ca. 1.5 cm long including the style. Occasional. Dry, rocky ridges, 6,500–7,000 ft. Red Fir F., Subalpine F. Dorleska Summit (*578*). June–July.

Rosa L.

Sepals pilose to glabrous on margins; prickles stout, flattened, and ± recurved . . . . *R. californica*

Sepals pectinate-glandular on margins; prickles slender, rounded, ± straight . . . . *R. nutkana*

R. californica Cham. & Schlecht. California Rose. Low shrub with prickly branches; stems up to 1.5 m high, bearing stoutish, flattened, ± recurved prickles; lvs. pinnate, the lfts. 5–7, obovate, 1–2 cm long, serrate, glabrous, the rachis and petiole ± pubescent; pedicels ca. 1 cm long, glabrous; sepals 8–10 mm long, caudate-attenuate, sparingly pubescent without, densely white-woolly within and along the margins; petals pink, 1.2–5 cm long; stamens many; pistils many, included within the fl.-tube; fr. globose to ovoid, 8–16 mm long, 10–15 mm thick. Found at one location. Stream bank, 6,800 ft. Red Fir F. Landers L. (*1019*). August.

R. nutkana Presl. Nootka Rose. Shrubs spreading from base, 1–2 m tall; stems bearing slender, rounded, ± straight prickles; lvs. pinnate, the lfts. 5–9, elliptic, 1–3 cm long, doubly serrate with gland-tipped teeth, rachis and petioles stipitate-glandular and bearing prickles; pedicels ca. 2 cm long, stipitate-glandular; sepals ca. 10 mm long, pectinate-glandular on the caudate tips, ± pubescent without, woolly within; petals pink, 2–3 cm long; stamens many; pistils many, included within the fl.-tube; fr. globose, 15–18 mm in diam. Found at one location. Stream bank, 5,400 ft. Mixed Conifer F. Canyon Cr. (*169*). June–July.

Rubus L.

Stems armed with prickles; lvs. pinnate . . . . *R. leucodermis*
Stems unarmed; lvs. simple . . . . *R. parviflorus*

R. leucodermis Dougl. ex T. & G. White-stemmed Raspberry. Erect to spreading woody vines armed with straight to hooked bristles that are ± flattened at the base; bark brownish, smooth; lvs. pinnate, lfts. 3, ovate to oblong, 1.5–4 cm long, doubly serrate, thinly pubescent above, densely white-tomentose beneath; infl. 3–10 fld.; sepals 8–12 mm long, acuminate, tomentose; petals white, narrow, ca. 5 mm long; stamens many; pistils many; fr. an aggregate of small druplets, dark purple to blackish or yellow-red, ca. 1.5 cm in diam. Found at one location. Wooded slope, 6,200 ft. Red Fir F. Caribou Mt. (*761*). July.

R. parviflorus Nutt. Thimble Berry. Low, branching shrub, 0.2–1 m tall; bark brown, peeling in age; lvs. simple, palmately lobed, cordate at base, serrate, 4–12 cm broad, dark green above, paler beneath, the blades thinly pubescent, petioles glandular; fls. several; sepals 8–14 mm long, attenuate, tomentose and glandular; petals white, broad, 10–15 mm long; stamens many; pistils many; fr. an aggregate of small drupelets, scarlet, 1–1.6 cm broad. Common. Woods, 5,000–6,000 ft. Mixed Conifer F., Red Fir F. Along the South Fork of the Salmon R. (*421*), Canyon Cr. (*834*). July–August.

Sibbaldia L.

S. *procumbens* L. Sibbaldia. Low, cespitose perennial from a branched, woody caudex; flowering stems slender, 4–10 cm high, strigose; lvs. pinnate, lfts. 3, obovate, coarsely toothed at the apex, 10–15 mm long, thinly pubescent, petioles slender, thinly strigose, 1–3 cm long; infl. cymose, compact, several fld.; sepals 2.5–3 mm long, thinly strigose, alternating with shorter, narrower bractlets; petals pale yellow, narrow, ca. 1.5 mm long; stamens 5; pistils ca. 10; achenes ca. 0.5 mm long; styles lateral. Common. Damp, open slopes, 7,000–8,000 ft. Subalpine F., Alpine Fell-fields. Canyon Cr. (*817*), below ice field on the north side of Thompson Pk. (*921*), Caribou Basin (*1050*). August.

Sorbus L.

S. *scopulina* Greene. Mountain Ash. Shrub, 1–3 m tall; bark brown to gray; lvs. pinnate, lfts. mostly 11–13, lanceolate, oblong to obovate, coarsely serrate, 2–5 cm long, glabrous, dark green above, paler beneath; infl. flat-topped, many fld.; sepals 1.5 mm long, glabrate without, pilose within; petals white, ca. 5 mm long; stamens 15–20; styles 4–5; fr. a globose pome, 8–10 mm in diam., orange to scarlet. Occasional. Wooded slopes, 5,000–6,000 ft. Mixed Conifer F. Josephine L. Basin (*595*). June–July.

Spiraea L.

Fls. in ± flat-topped cymes; sepals remaining erect....... S. *densiflora*
Fls. in elongate panicles; sepals soon reflexed............. S. *douglasii*

S. *densiflora* Nutt. ex T. & G. Rose-colored Meadow-sweet. Deciduous shrub 2–9 dm tall; bark reddish brown, peeling in age; lvs. oblanceolate to obovate, serrate mostly above the middle, glabrous, 1–4 cm long; infl. a densely fld., ± flat-topped cyme; sepals 0.5 mm long, ± pubescent at tips, remaining erect; petals rose to purple, ca. 1.5 mm long; stamens many; pistils mostly 5; follicles 2–2.5 mm long, ± shining. Common. Moist, rocky places, 6,000–7,000 ft. Red Fir F. Caribou Basin (*731*), Lander L. (*1015*). July–August.

S. *douglasii* Hook. Douglas' Spiraea. Deciduous shrubs, 1–2 m tall; bark tan to dark reddish brown, peeling in age; lvs. elliptic to oblong, coarsely serrate above the middle, glabrous above, sparingly to densely tomentose below, 2–6 cm long; infl. a densely fld., elongate panicle; sepals ca. 1 mm long, tomentose, soon reflexed; petals rose to purple, ca. 1.5 mm long; stamens many; pistils mostly 5; follicles ca. 2.5 mm long, brown, shining. Common. Meadows and stream banks, 5,000–6,000 ft. Mixed Conifer F. Big Flat (*220*), head of South Fork of the Salmon R. (*422*), Boulder Cr. L. (*471*), Canyon Cr. Meadow (*855*). August.

RUBIACEAE

Lvs. opposite; corolla pink . . . . *Kelloggia*
Lvs. 2–6 in a whorl; corolla purplish red or greenish white . . . . . . *Galium*

Galium L.

Lvs. mostly 6 in a whorl . . . . *G. triflorum*
Lvs. mostly 2–4 in a whorl.
  Fr. glabrous . . . . *G. bolanderi*
  Fr. bearing bristles or hairs.
    Fr. bearing hooked bristles; stems slender . . . . *G. bifolium*
    Fr. bearing long, tawny hairs; stems stout, from a woody base.
    G. glabrescens

G. bifolium Wats. Low Mountain Bedstraw. Glabrous, slender-stemmed annual; stems 5–10 cm tall; lvs. 2–4 in a whorl, oblanceolate to linear, 5–15 mm long; pedicels axillary and terminal, 1-fld., recurved and equaling lvs. in fr.; fls. whitish, ca. 1 mm wide; fr. ca. 2 mm in diam., bearing stiff, hooked hairs. Occasional. Damp slopes and meadows, 6,000–7,000 ft. Red Fir F. Bullards Basin (*636*), Kidd Cr. (*698a*). June–July.

G. bolanderi Gray. Bolander's Bedstraw. Perennial from basal woody stems and a central taproot; stems mostly erect, 10–25 cm tall; lvs. 4 in a whorl, oblanceolate, 5–15 mm long, glabrous on the surfaces, hispid-ciliate on margins; male fls. in small terminal cymes, female fls. solitary in upper axils; pedicels short, erect to recurved; fls. purplish red, ca. 2 mm in diam.; fr. 3–3.5 mm in diam., glabrous. Found at one location. Dry, rocky slopes, 6,500 ft. Red Fir F. Packers Pk. (*939*). July–August.

G. glabrescens (Ehr.) Dempst. & Ehr. Gray's Bedstraw. Fig. 71. Perennial from a stout, woody caudex; stems 5–20 cm tall, stiffly puberulent; lvs. 4 in a whorl, ovate, 5–15 mm long; plants dioecious; infl. narrow, leafy, 3–10 cm long; fls. greenish white, the corolla lobes hispid-puberulent externally; fr. ca. 3 mm in diam., covered with tawny hairs ca. 3 mm long. Common. Dry, rocky slopes, 6,000–8,000 ft. Red Fir F., Subalpine F. North side of Red Rock Mt. (*639*), upper end of Canyon Cr. (*814*). July–August. The nomenclature here is that of Dempster and Ehrendorfer (1965).

G. triflorum Michx. Fragrant Bedstraw. Low, trailing perennial; stems 20–80 cm long, ± scabrous; lvs. mostly 6 in a whorl, oblanceolate to obovate, thin, 10–50 mm long, surface glabrous, margins and midribs minutely scabrous; fls. mostly in threes in axillary cymules, pedicels long and slender, the peduncles bearing a whorl of leafy bracts; fls. greenish white, ca. 2 mm wide; fr. ca. 2 mm wide, densely bristly. Common. Shaded stream banks and woods, 5,000–6,000 ft. Mixed Conifer F. Along South Fork of Salmon R. (*406, 1033*), Canyon Cr. (*831*). July–August.

Kelloggia Torr.

K. galioides Torr. Kelloggia. Perennial from slender rootstocks; stems wiry, glabrous, 10–25 cm tall; lvs. opposite, lanceolate, 1–3 cm long; fls. several in loose, forking, terminal cymes; pedicels long and slender; corolla pink, 3–4 mm long; fr. 3–4 mm long, covered with hooked bristles. Common. Dry, shaded woods, 5,000–6,000 ft. Mixed Conifer F. Big Flat (*362*), Kidd Cr. (*677*), Canyon Cr. (*784*). July–August.

SALICACEAE

Lvs. ovate to cordate . . . . *Populus*
Lvs. oblanceolate to elliptic . . . . *Salix*

Populus L.

P. tremuloides Michx. American Aspen. Tree to 10 m tall; bark gray, furrowed and cracked; lvs. ovate, rounded to cordate at base, crenulate or serrulate, 3–7 cm long, 2.5–5 cm wide, dark green above, paler beneath, petioles slender, 3–4 cm long; catkins 3–6 cm long; bracts 3–5 lobed, fringed with long hairs; capsule 5–7 mm long. Occasional. Stream banks and wet meadows, 5,000–6,000 ft. Mixed Conifer F. Bullards Basin (*631*), Boulder Cr. Basin (*1432*). June–July.

Salix L.

Lvs. mostly less than 10 mm wide, sparingly pubescent on lower surface; catkin scales yellow . . . . *S. exigua*
Lvs. mostly 10–25 mm wide, silky-pubescent to tomentose on lower surface; catkin scales brown to very dark.
  Capsule glabrous; scales very dark . . . . *S. commutata*
  Capsule densely silky; scales brown . . . . *S. jepsonii*

S. commutata Bebb. Sierra Willow. Low shrub to 3 m high; bark reddish brown, densely puberulent on young twigs; lvs. oblanceolate to elliptic, 1–7 cm long, 10–25 mm wide, silky-pubescent; female catkins on leafy peduncles; scales oblanceolate, dark, 1.5–2 mm long, villous; stamens 2, filaments glabrous; capsule glabrous, 6–7 mm long. Occasional. Stream banks, 5,000–7,300 ft. Mixed Conifer F., Red Fir F., Subalpine F. Canyon Cr. (*801, 1314*). July–August.

S. exigua Nutt. Slender Willow. Low, spreading shrub to 1 dm high; lvs. linear to linear-lanceolate, 2–6 cm long, 3–6 mm wide, sparingly pubescent on both surfaces, serrulate; scales yellow, ca. 1.5 mm long, villous below; stamens 2, filaments villous-woolly below; capsule glabrous, ca. 5 mm long. Found at one location. Edge of lake, 7,100 ft. Red Fir F. Landers L. (*1468*). July–August.

S. jepsonii C. K. Shneid. Jepson's Willow. Shrub to small tree 1–3 m tall;

bark brown to purple, ± pubescent on young twigs; lvs. oblanceolate, 3–7 cm long, 10–20 mm wide, green and ± silky-pubescent above, whitish and silky-tomentose below; female catkins short pedunculate; scales brown, oblong, ca. 1.5 mm long, pilose; stamens 2, filaments glabrous; capsule densely silky, 4.5–5.5 mm long. Found at one location. Stream bank, 5,800 ft. Mixed Conifer F. Kidd Cr. (*683*). July–August.

SARRACENIACEAE

Darlingtonia Torr.

D. californica Torr. California Pitcher-plant. Fig. 72. Scapose perennial from short, fibrous rootstocks; stems 2.5–5 cm tall, bearing scattered, lanceolate bracts 2–3 cm long; lvs. basal, 7–30 cm long, tubular, expanded into a hood at the summit and bearing a 2-lobed appendage at the apex; fls. solitary, pendulous; sepals yellowish green, 3–4.5 cm long; petals dark purple-red, veiny, 2.5–3 cm long; stamens 12–15; capsule obovoid, 2.5–4.5 cm long. Common. Bogs, 5,000–6,800 ft. Mixed Conifer F., Red Fir F. Mouth of Bullards Basin (*627*), head of Union Cr. (*648*), head of Sunrise Cr. (*667*), Landers L. (*1017*). July–August.

SAXIFRAGACEAE

Plants woody shrubs; fruit a berry . . . . *Ribes*
Plants herbaceous, not woody; fruit a capsule or follicle.
  Fertile stamens 10.
    Styles 3; petals laciniate or toothed. . . . . *Lithophragma*
    Styles 2.
      Petals entire, white, some with yellow spots. . . . *Saxifraga*
      Petals laciniate-pinnatifid, whitish to red . . . . *Tellima*
  Fertile stamens 5.
    Fertile stamens alternating with clusters of gland-tipped staminodia . . . . *Parnassia*
    Fertile stamens not alternating with staminodia.
      Petals cleft . . . . *Mitella*
      Petals entire.
        Stems from short, bulbiferous rootstocks. . *Suksdorfia*
        Stems from scaly rhizomes.
          Lvs. 10–20 cm broad; petals broadly oval to obovate, 4–5 mm long . . . . *Boykinia*
          Lvs. 2–8 cm broad; petals narrow, 2 mm long. *Heuchera*

Boykinia Nutt.

B. major Gray. Mountain Boykinia. Fig. 73. Aromatic perennial from thick, woody rhizomes; stems simple, 3–9 dm high, 4–7 mm thick at base, glandular-pubescent; lvs. alternate, the lower ones on petioles 1–2 dm long, the blades round-reniform in outline, 1–1.5 dm broad, cleft into 5–7 lobes which

Cercocarpus ledifolius

Luetkea pectinata

Galium glabrescens

Darlingtonia californica

are again cleft and tipped with glandular teeth; infl. dense, many-fld., densely glandular-pubescent; fl.-tube 2–3 mm long; sepals triangular lanceolate, acute; petals white, 5–7 mm long; base of pistil surrounded by a golden yellow nectary. Common. Damp, rocky areas along streams and lakes, 5,600–7,600 ft. Mixed Conifer F., Red Fir F. Josephine L. (*289*), Caribou Basin (*395*, *738*), Canyon Cr. (*778*), Boulder Cr. (*472*). July–August.

Heuchera L.

Fls. rounded at base, 2–4 mm long; flowering stems 1–3 dm tall. *H. merriamii*

Fls. turbinate, 1–3 mm long; flowering stems 3–7 dm tall. *H. micrantha* var. *pacifica*

H. merriamii Eastw. Trinity Alum Root. Perennials from woody rhizomes; lvs. basal, rounded in outline, cordate to truncate at base, 5–8 lobed with narrow, deep sinuses, blades 2–5 cm broad on petioles 2–7 cm long; flowering stems scapose, hirsute; panicle narrow, ± compact, glandular, 5–15 cm long; fls. 3–4 cm long on slender pedicels; sepals pale green to cream, sometimes rose-tinted, hirsute; petals white, narrow, spatulate, ca. 1.5 mm long; styles rather slender, exserted as far as stamens in age. Common. Rocky ledges and outcrops, 5,500–7,500 ft. Mixed Conifer F., Red Fir F., Subalpine F. Josephine L. (*82*, *594*), Black Mtn. (*454*, *1093*), Caribou Basin (*1061*), Grizzly L. (*890*), Sawtooth Ridge (*1067*). June–August.

H. micrantha Dougl. ex Lindl. var. *pacifica* Rosend., Butt., & Lak. Small-flowered Heuchera. Coarse perennial from stout caudex; lvs. basal, 5–7 lobed, the lobes with 3–5 broad, ± cuspidate teeth separated by broader sinuses, blades 3–8 dm long, on petioles 5–18 cm long; flowering stems scapose, 3–7 dm tall; panicle narrow and somewhat lax; fls. 1–3 mm long, on slender pedicels; sepals greenish; petals white, oblanceolate, with narrow claws, ca. 2 mm long; stamens and styles exserted. Seen on a single rock ledge, 5,000 ft. Mixed Conifer F. Canyon Cr. (*65*). June.

Lithophragma Nutt.

Basal lvs. shallowly lobed . . . . *L. heterophylla*

Basal lvs. lobed almost to their bases.

Stem lvs. often bearing axillary bulblets which replace the fls. *L. bulbifera*

Stem lvs. not bearing axillary bulblets. . . . . *L. parviflora*

L. bulbifera Rydb. Rock Star. Stems from slender bulblet-bearing rootstocks, glandular-pubescent, 1.5–2 dm tall; basal lvs. deeply lobed, stem lvs. often bearing bulblets; fls. 3–8; pedicels 2–5 mm long; fl.-tube rounded at base, densely glandular-pubescent, 2–3 mm long; petals white to pink, divided into narrow lobes, narrowly clawed. Occasional. Open meadows, 6,500 ft. Red Fir F. Bullards Basin (*633a*), Kidd Cr. (*695*). July.

L. heterophylla (H. & A.) T. & G. Hill Star. Stems simple, 2.5–5 dm high; basal lvs. crenately lobed to cleft half way to base; 1.5–4 cm wide; stem lvs. alternate, more deeply lobed; fls. 3–9; fl.-tube 2–4 mm high and wide rounded to truncate at base with evident parallel veins; sepals 2 mm long; petals white, toothed to deeply lobed. Common. Dry slopes and openings in woods, 5,000–6,500 ft. Mixed Conifer F., Red Fir F. Canyon Cr. (*66*), Big Flat (*518*), Packers Pk. (*555*). June.

L. parviflora (Hook.) Nutt. Prairie Star. Stems from slender bulblet-bearing rootstocks, simple, scapose, glandular-pubescent, 2–3.5 dm tall; basal lvs. 2–3 cm wide with 3–5 divisions which are in turn lobed; petioles and lvs. covered with stiff whitish hairs; fl.-tube elongate-obconic, 3–4 mm long with a pale, horizontal band at summit; sepals deltoid, 1–1.5 mm long; petals white or pinkish, 5–10 mm long, deeply 3–5 cleft. Seen at one location. Open woods, 5,000 ft. Mixed Conifer F. Big Flat (*519*). June. In California this species has been previously reported from the Sierra Nevada and the mts. of southern California (Munz, 1959).

Mitella L.

Petals green, pinnately cleft. . . . . *M. pentandra*
Petals white, palmately 3-cleft at apex. *M. trifida*

M. pentandra Hook. Alpine Mitrewort. Delicate perennial from slender rootstocks; scapes slender, glabrous, leafless, 1–3 dm tall; basal lvs. cordate, round-reniform in outline, crenate to serrate, 2–6 cm wide on slender petioles 2–10 cm long; fls. 8–20; fl.-tube ca. 3 mm wide; sepals greenish; petals with 5–7 filiform lobes. Common. Damp, shaded slopes and banks, 5,500–7,300 ft. Red Fir F. Kidd Cr. (*686*), Caribou Basin (*1062*), Grizzly Cr. (*865*). July–August.

M. trifida Grah. Pacific Mitella. Perennial herbs from relatively long rootstocks with many fibrous roots; scapes slender, pubescent, 1–3 dm tall; lvs. basal, cordate to round-reniform with stiff hairs on blades and petioles, blades 2–6 cm long; fls. 5–25; fl.-tube ca. 2 mm long, puberulent; sepals whitish; petals ca. 2 mm long, narrow, 3-cleft near apex. Occasional. Shaded banks and open woods, 5,000–6,000 ft. Mixed Conifer F. Big Flat (*93*), Yellow Rose Mine Tr. (*570*). June.

Parnassia L.

Petals entire, 10–18 mm long. . . . . *P. palustris* var. *californica*
Petals fimbriate on the sides of basal half, mostly 10–12 mm long.
P. fimbriata

P. fimbriata Konig. Fringed Grass-of-Parnassus. Glabrous plants from short rootstocks; stems simple, 2–5 dm high with one sessile bract above the middle; lvs. basal, reniform to ovate, 2–4 cm broad; fl. solitary; sepals elliptic to oval,

5–6 mm long; petals white, clawed; staminodia united into lobed scales. Occasional. Damp, shaded slopes, 6,200–7,800 ft. Red Fir F. Kidd Cr. (*435*), Caribou Basin (*1060*). August. The material from Caribou Basin has staminodia divided into filiform segments ca. 2 mm long. This is characteristic of *P. intermedia* and *P. cirrata* but the plants seem closest to *P. fimbriata* in other respects, e.g., petals, sepals, and lvs.

P. palustris L. var. *californica* Gray. California Grass-of-Parnassus. Fig. 74. Glabrous plants from short rootstocks; lvs. basal, ovate, 2.5–4 cm long, cuneate at base; petioles 2–10 cm long; scapes 2.5–5 dm tall with a sessile bract above the middle; fl. solitary; sepals 4–6 mm long; petals white, oblong-ovate to roundish; staminodia in groups of 15–20, capillary, gland-tipped. Occasional. Wet, shaded meadows, 6,000 ft. Red Fir F. Kidd Cr. (*434*). August.

Ribes L.

Nodal spines lacking.
  Fls. greenish white . . . . *R. cereum*
  Fls. rose to deep red.
    Sepals erect; lvs. glabrous above, ± pubescent beneath. *R. nevadense*
    Sepals spreading or recurved; lvs. puberulent above, white pubescent to tomentulose beneath . . . . *R. sanguineum*
Nodal spines present.
  Free part of fl.-tube saucer-shaped . . . . *R. lacustre*
  Free part of fl.-tube campanulate to cylindric.
    Anthers blunt at apex, almost as broad as long.
      Sepals 4–5 mm long; lvs. not glandular beneath. *R. binominatum*
      Sepals 10–12 mm long; lvs. glandular beneath . . *R. lobbii*
    Anthers apiculate, much longer than broad . . . . *R. roezlii*

R. binominatum Heller. Trailing Gooseberry. Low, trailing plants with branches to 1 m long; nodal spines mostly 3, to 2 cm long; lvs. roundish in outline, deeply 3–5 lobed, cordate, ± paler beneath; fls. greenish white on hairy peduncles, fl.-tube ca. 4 mm long; sepals 4–5 mm long, pubescent; petals 2–4 mm long; stamens exceeding petals; berry ca. 1 cm in diam., covered with stiff spines up to 1 cm long. Occasional. Trailing on forest floor, steep, wooded slopes, 5,000–7,500 ft. Mixed Conifer F., Red Fir F. Caribou Basin (*745*), Big Flat (*1143*). June–July.

R. cereum Dougl. White Squaw Currant. Fragrant shrub, erect, much branched, 1–12 dm tall; lvs. round-reniform in outline, 3–5 lobed, crenulate, puberulent, and glandular, 1–5 cm broad on petioles 4–5 cm long; fl.-tube 6–8 mm long, greenish white; sepals rounded, 1–3 mm long; petals white, rounded; berry red, slightly glandular-hairy, ca. 6 mm in diam. Occasional. Dry, rocky slopes, 6,000–7,000 ft. Red Fir F. Caribou Basin (*187*), Sunrise Cr. (*666*). July.

R. lacustre (Pers.) Poir. Swamp Currant. Prostrate shrubs, the branches to 1 m long and ± trailing; nodal spines 3–5, internodes bristly; lf.–blades 3–5 cm wide, deeply 3–5 lobed, round-cordate in outline, glabrous; petioles slender, 2–5 cm long; racemes many, loosely spreading; fl.-tube saucer-shaped above the ovary, purplish, glandular-bristly; sepals short, broad, 1.5 mm long; petals smaller; stamens included; berry black-purple with weak glandular hairs, ca. 4 mm in diam. Occasional. Wet, open slopes, 6,000 ft. Mixed Conifer F., Red Fir F. Yellow Rose Mine Tr. (*571*). June.

R. lobbii Gray. Gummy Gooseberry. Shrubs with stems 1–2 m tall, young twigs glandular-pubescent; nodal spines mostly 3, stout, to 1.5 cm long; lvs. roundish in outline, 1.5–5 cm wide, 3–5 lobed, crenately toothed, ± glabrous above, glandular-pubescent beneath; fl.-tube 3–4 mm long, purple; sepals purple, reflexed, 10–12 mm long; petals pale yellow to white, involute, ca. 4 mm long; berry oblong, densely glandular-bristly, 12–15 mm long. Seen at one location. Dry, wooded slope, 5,000 ft. Mixed Conifer F. Big Flat (*526*). June.

R. nevadense Kell. Sierra Nevada Currant. Openly branched shrubs, 1–2 m tall, glabrous to puberulent on young growth; lvs. roundish in outline, 3–5 lobed, crenate, 3–7 cm wide, resinous-dotted; racemes 8–12 fld., spreading or drooping; fls. rose to deep red, free part of tube ca. 2 mm long; sepals 2 mm long, erect; petals white, shorter; berry blue-black, ± glandular, ca. 8 mm in diam. Common. Shaded woods and stream banks, 5,800–6,800 ft. Mixed Conifer F., Red Fir F. Union L. (*640*), Kidd Cr. (*688*), Caribou Basin (*734*), "L" Lake (*832*). July–August.

R. roezlii Regel. Sierra Gooseberry. Fig. 75. Stout shrubs, 5–12 dm high with spreading branches; nodal spines 1–3; lvs. roundish in outline, 1–2.5 cm wide, 3–5 cleft into toothed lobes; free portion of fl.-tube ca. 6 mm long, purplish; sepals 7–10 mm long, purplish red, pubescent; petals whitish, 3–5 mm long; filaments ca. as long as petals; berry purple or lighter, covered with long spines, 14–16 mm in diam. Common. Dry woods and open slopes, 5,000–6,500 ft. Mixed Conifer F., Red Fir F. Canyon Cr. (*62*, *859*), Big Flat (*501*), Yellow Rose Mine (*572*). June–August.

R. sanguineum Pursh. Red Flowering Currant. Erect shrubs, 1–3 m tall, young growth puberulent; lvs. round-reniform in outline, 3–5 lobed, serrulate, dark green and puberulent above, paler beneath and white puberulent to tomentulose, 2–7 cm wide; racemes 10–20 fld. exceeding lvs.; bracts 6–8 mm long, red or purplish; fls. deep rose, free part of tube ca. 4 mm long; sepals 4–5 mm long; petals ca. half as long, red or paler; berry black, glandular, roundish to oblong. Seen at one location. Damp, wooded slope, 5,000 ft. Mixed Conifer F. Big Flat (*538*). June.

Saxifraga L.

Lf. blades rounded in outline.
  Lf. blades simply and evenly dentate; petioles without stipules. *S. punctata* ssp. *arguta*
  Lf. blades with large teeth that are usually 3-toothed at apex; petioles with stipular dilations .... *S. mertensiana*
Lf. blades not rounded, usually longer than broad.
  Lower stems ± prostrate, covered with linear lvs......... *S. tolmiei*
  Lower stems not as above; lvs. mostly basal.
    Petioles narrow, jointed with the fan-shaped lf. blades. *S. fragarioides*
    Petioles not jointed with lf. blades.
      Petals unlike, 3 broader than the other 2; infl. spreading.
        Stems 0.8–1.3 dm high; fls. few, mostly terminal. *S. bryophora*
        Stems 1–3 dm high; fls. more numerous. *S. ferruginea*
      Petals all alike; infl. more compact.
        Scapes glabrous at top, becoming glandular-pubescent toward base, 0.3–1.5 dm tall .... *S. aprica*
        Scapes glandular-pubescent throughout, 1–3 dm tall. *S. nidifica*

S. aprica Greene. Sierra Saxifrage. Plants usually purplish; lvs. in basal tuft, ovate to oblong or spatulate, 1–4 cm long; infl. subcapitate; sepals 1–1.5 mm long; petals white, 1.5–2 mm long; stamens shorter than petals; capsule 2.5–4 mm long, with divergent tips. Occasional. Damp, rocky meadows and seeps, 7,000 ft. Red Fir F., Subalpine F. Landers L. (*660*), Grizzly L. (*892*). July–August.

S. bryophora Gray. Bud Saxifrage. Slender stemmed annual 0.8–1.3 dm high, glandular-pubescent, with a few to many branches; lvs. in basal tufts, oblanceolate, 1–2.5 cm long, denticulate toward apex, ciliate; all but uppermost fls. bulbiferous, the pedicels capillary, 2–3 mm long, spreading to reflexed in age; sepals 1–1.5 mm long, linear to ovate, reddish, reflexed; petals unequal, white with 2 yellow spots, 2.5–4 mm long including claw; capsule ca. 3 mm long. Occasional. Wet, rocky places, 5,900–6,700 ft. Mixed Conifer F., Red Fir F. Boulder Cr. Basin, (*1289*), Caribou Basin (*I. L. Wiggins 13,547*). July. In California this species has been previously reported only from the Sierra Nevada (Munz, 1959).

S. ferruginea Grah. Alaska Saxifrage. Stems slender, 1–3 dm high, openly and paniculately branched, some of the fls. replaced by axillary bulblets; lvs. in basal tuft, spatulate to ± oblanceolate, 1.5–4 cm long, sharply toothed and ± ciliate; sepals reflexed, 1–2 mm long; petals white, unequal, 3 lanceolate and clawed, with 2 yellow basal spots, 2 smaller, spatulate and usually lacking basal spots; stamens filiform; capsule 2.5–4 mm long with divergent tips. Common. Wet meadows and seeps, 6,000–7,000 ft. Red Fir F., Subalpine F. Josephine L. (*283*), Kidd Cr. (*447, 1106*), Caribou Basin (*729*), Grizzly L. (*893*). July–August.

S. *fragarioides* Greene. Joint-leaved Saxifrage. Coarse herbs from woody rootcrown with stout, horizontal branches; lvs. in basal tufts, cuneate-obovate, abruptly contracted to and jointed with petioles, blades 1.5–4 cm long, dentate at summit; flowering stems 1–2.5 dm tall, with bractlike lvs., bearing narrow, terminal panicles; sepals reflexed in age, 1.5–2 mm long; petals white, 2–3 mm long, reflexed in age; capsule ca. 4–5 mm long. Common. Rock ledges and walls, 6,500–7,700 ft. Red Fir F., Subalpine F. Caribou Mtn. (*347*), Caribou Basin (*730*), Canyon Cr. (*816*), Sawtooth Ridge (*1054*). July–August.

S. *mertensiana* Bong. Wood Saxifrage. Perennial from scaly, bulblike rootstocks; lvs. basal, round, 2–7 cm wide, main teeth dentate; flowering stems 1–3 dm tall, finely glandular-pubescent; infl. paniculate, open; sepals 1.5–2.5 mm long, reflexed; petals white to pink, 3–5 mm long; filaments dilated; capsule 5–7 mm long. Seen at one location. Wet, rocky cliffs, 6,700 ft. Red Fir F. Caribou Basin (*755*). July.

S. *nidifica* Greene. Peak Saxifrage. Scapose herbs from a short, bulbiferous caudex; lvs. in basal tufts, 1–4 cm long on shorter petioles; flowering stems 1–3 dm tall, glandular-pubescent throughout; infl. narrowly paniculate; sepals 1.5 mm long; petals white, ca. 2 mm long; filaments ± dilated toward base; capsule 2–2.5 mm long, the tips ± divergent. Common. Damp meadows and rocky places, 5,000–6,500 ft. Mixed Conifer F., Red Fir F. Josephine L. (*73*), Big Flat (*509*), Kidd Cr. (*696*). June–July.

S. *punctata* L. ssp. *arguta* (D. Don) Hult. Brook Saxifrage. Glabrous perennial from a rhizome; lvs. basal, round, 1.5–7 cm wide, glabrous, coarsely dentate; petioles 2–20 cm long; scapes slender, 2–4 dm tall; infl. open, spreading; sepals 1.5–2 mm long, reflexed, usually purplish; petals white with 2 yellow spots, 2.5–5 mm long with distinct claws; filaments broadened upwards; follicles 6–8 mm long, purplish. Common. Damp, shaded stream banks, 6,500–7,300 ft. Red Fir F. Grizzly Cr. (*902*), Caribou Basin (*1059*), Kidd Cr. (*1105*). August.

S. *tolmiei* T. & G. Alpine Saxifrage. Densely tufted, diffusely branched perennials, the stems ± prostrate; lvs. succulent, glabrous, closely imbricated, 0.8–1.5 cm long; flowering stems slender, 0.3–1.5 dm tall, often reddish, few-branched at summit; sepals 2 mm long; petals white, elliptic-spatulate, 4–5 mm long; filaments dilated; capsule 8–12 mm long, purplish. Common. Wet, rocky slopes, 7,400–8,500 ft. Subalpine F., Alpine Fell-fields. Caribou Mt.

(*356*), Grizzly L. (*916*), Caribou Basin (*1053*). August.

Suksdorfia Gray

S. ranunculifolia (Hook.) Engler. Buttercup-leaved Suksdorfia. Plants from slender rootstocks bearing reddish bulblets; stems 1–3 dm tall, glandular-puberulent; basal lvs. ± reniform in outline, crenate, ternately divided, 1–2 cm long; petioles 2–10 cm long; upper lvs. smaller; infl. a short, compact cyme; sepals triangular-ovate, 1.5–2 mm long; petals white, 4–6 mm long. Common. Wet, rocky places, 5,500–6,600 ft. Red Fir F. Josephine L. (*70, 593*), Canyon Cr. (*158*), Caribou Basin (*200*). June–July.

Tellima R. Br.

T. grandiflora (Pursh) Dougl. Fringe Cups. Perennials from stout rootstocks; flowering stems 4–8 dm tall, spreading hirsute; basal lvs. roundish in outline, ± hirsute, shallowly 3–7 lobed, serrate to crenate, 5–10 cm broad, on petioles 5–20 cm long; racemes few to many fld.; pedicels 1–3 mm long; fl.-tube glandular-puberulent, striate, 3–5 mm long; sepals 1.5–2 mm long; petals at first whitish, later red, 4–6 mm long, the segments filiform. Occasional. Woods and rocky places, 5,500–7,000 ft. Red Fir F. Josephine L. (*599*), Canyon Cr. (*789*), Grizzly L. Basin (*862, 885*). June–August.

SCROPHULARIACEAE

Stigmas broad, flattened, distinct . . . . *Mimulus*
Stigmas capitate, wholly united.
  Corolla with upper lip flattened or widely arched, not forming a galea which encloses the stamens.
    Fertile stamens 2; upper lip of corolla appearing 1-lobed. *Veronica*
    Fertile stamens 4; upper lip of corolla 2-lobed.
      Middle lobe of the lower lip of the corolla keel-shaped and enclosing stamens . . . . *Collinsia*
      Middle lobe of the lower lip of the corolla broad and not enclosing stamens . . . . *Penstemon*
  Corolla with upper lip narrowly arched, forming a galea which encloses the stamens.
    Anther cells equal in size and position . . . . *Pedicularis*
    Anther cells unequally placed.
      Calyx divided laterally, completely surrounding the lower portion of corolla.
        Lower lip of corolla larger than galea; corollas rose to lavender . . . . *Orthocarpus*
        Lower lip of corolla smaller than galea; corollas reddish orange to scarlet or yellow. . . . . *Castilleja*
      Calyx divided ventrally completely surrounding corolla only at very base, if at all. . . . . *Cordylanthus*

Castilleja Mutis.

Corollas cream to lemon yellow; plants arachnoid-woolly. *C. arachnoidea*

Corollas reddish orange to scarlet; plants glandular-pubescent to mostly glabrous.

Plants glandular-pubescent below infl. . . . . *C. applegatei*

Plants glabrous below infl. . . . . *C. miniata*

C. applegatei Fern. Wavy-leaved Indian Paint-brush. Glandular-pubescent perennial from a woody base; stems clustered, simple to branched, 2–5 dm tall; lvs. glandular-pubescent, lanceolate, sessile, 2–3.5 cm long, entire to 3-lobed with narrow segments; bracts and calyces distally reddish orange to scarlet; calyx 12–22 mm long, divided laterally into lanceolate lobes 2–4 mm long; corolla 20–30 mm long, the galea reddish, 12–15 mm long, the lower lip greenish, 1–2 mm long. Common. Dry, open slopes and flats, 5,000–7,500 ft. Mixed Conifer F., Red Fir F., Subalpine F. Big Flat (*105, 540*), Packers Pk. (*560*), Dorleska Mine (*626*), Bullards Basin (*637*), Caribou Basin (*746, 751*), Caribou Mt. (*719*), Kidd Cr. (*684*), north side of Thompson Pk. (*910*). June–August.

C. arachnoidea Greene. Cobwebby Indian Paint-brush. White-woolly perennial from a woody rootcrown; stems many, clustered, 1–2.5 dm tall; lvs. pubescent to woolly, lance-linear, 2–4 cm long including petioles, entire to 2–4 lobed, the lobes narrow and spreading; infl. 2–10 cm long; bracts and calyces greenish; calyx 12–14 mm long, cleft into lance-linear lobes; corolla yellow, 10–12 mm long, the galea 3–4 mm long, the lower lip slightly shorter. Common. Open, dry slopes, 7,000–8,100 ft. Red Fir F., Subalpine F. Caribou Basin (*749*), ridge south of Thompson Pk. (*812*), below ice field on north side of Thompson Pk. (*924*), Red Rock Mt. (*1027*), Caribou Mt. (*1040*). July–August.

C. miniata Dougl. ex Hook. Great Red Indian Paint-brush. Erect perennial from a woody rootcrown; stems mostly simple, glabrous below the infl., 3–8 dm tall; lvs. sessile, linear to narrowly elliptic, 2–5 cm long, acute, glabrous; infl. villous-pubescent, bracts and calyces distally reddish; calyx 20–27 mm long, laterally cleft into narrow lobes 3–7 mm long; corolla 25–35 mm long, the galea green with red margins, 15–20 mm long, the lower lip green, 1–2 mm long. Common. Damp meadows, slopes, and stream banks, 5,000–7,000 ft. Mixed Conifer F., Red Fir F. Big Flat (*232*), Boulder Cr. Basin (*463*), Caribou Basin (*764*), Canyon Cr. (*787*), Landers L. (*1014*), Kidd Cr. (*1112*). July–August.

Collinsia Nutt.

Infl. not glandular . . . . *C. parviflora*

Infl. glandular-pubescent.

Fruiting pedicels ascending. . . . . *C. linearis*
Fruiting pedicels deflexed . . . . *C. torreyi* var. *latifolia*

C. linearis Gray. Rattan's Blue-eyed Mary. Low annual; stems simple, 10–20 cm tall; lvs. linear-lanceolate to ± spatulate, 1–3 cm long; infl. glandular-pubescent with linear bracts 2–4 mm long; calyx 3–4 mm long with lanceolate lobes; corolla 7–10 mm long, purple, the upper lip whitish to cream, ± reflexed. Found at one location. Dry, open slope, 7,000 ft. Red Fir F. Packers Pk. (*546*). June.

C. parviflora Dougl. ex Lindl. Small-flowered Blue-eyed Mary. Low annual; stems 5–15 cm tall, puberulent; lower lvs. oblanceolate, the upper becoming oblong, 1–2 cm long; infl. somewhat congested, fls. 1–2 in axils of bracts; calyx 4–6 mm long with narrow, acuminate lobes, scarious to purplish between the lobes; corolla 4–6 mm long, the upper lip whitish to blue at the tips, the lower lip purple at the tips, the tube pale. Common. Open meadows and slopes, 5,000 ft. Mixed Conifer F. Big Flat (*512*). June.

C. torreyi Gray var. *latifolia* Newsom. Torrey's Blue-eyed Mary. Low annual; stems simple to branched, 5–20 cm tall; lvs. lance-ovate to elliptic, 1–3 cm long, sessile or short petioled; infl. lax, ± glandular-pubescent, the upper bracts becoming reduced; calyx 2–4 mm long, the lobes narrow and acute; corolla 6–9 mm long, the upper lip whitish to cream with purple dots towards base, the lower lip blue. Common. Open meadows and slopes, 5,000–6,600 ft. Mixed Conifer F., Red Fir F. Big Flat (*99, 543*), Caribou Basin (*193*), Caribou Mt. (*217*). June–July.

Cordylanthus Nutt. ex Benth.

C. viscidus (Howell) Penn. Viscid Bird's-beak. Glandular-viscid annuals; stems few to many branched, 10–20 cm high; lvs. 2–3 cm long, divided into filiform segments; fls. borne singly or in groups of 2–3, each fl. subtended by 2–3 trilobed bracts plus a lanceolate bract resembling the calyx; calyx divided ventrally to base, entire at apex, glandular-pubescent, 10–15 mm long; corolla 12–15 mm long, yellow-green with reddish stripes, galea pubescent toward tip, reddish at base; filaments hairy on one side; capsule 7–8 mm long. Found at one location. Dry, sandy places, 5,200 ft. Mixed Conifer F. Boulder Cr. Basin (*1275*). July–August.

Mimulus L.

Corollas yellow.
Mature calyx strongly inflated, the lower lobes curved up against the others.
Corollas 1–1.5 cm long. . . . . *M. guttatus* ssp. *micranthus*
Corollas 1.5–4 cm long.
Lvs. slimy, ± glandular-puberulent; stems 1–2 dm tall. *M. tilingii*

Lvs. mostly glabrous; stems 1–5 dm tall. . . . . *M. guttatus*

Mature calyx not strongly inflated, the lobes remaining practically straight.

Corollas 8–10 mm long . . . . *M. pulsiferae*

Corollas 15–28 mm long.

Stems 5–30 cm long; plants ± slimy.

Plants densely pilose . . . . *M. moschatus*

Plants glabrous to finely pubescent.

M. moschatus var. *moniliformis*

Stems 1–5 cm long; plants not slimy.

Lvs. oblong to obovate. . . . . *M. primuloides*

Lvs. linear-oblanceolate.

M. primuloides var. *linearifolius*

Corollas rose to purple.

Corollas 28–40 mm long . . . . *M. lewisii*

Corollas 6–20 mm long.

Corollas 6–10 mm long . . . . *M. breweri*

Corollas 12–20 mm long.

Stigmas and long stamens exserted from corolla throat.

M. nanus

Stigmas and stamens included in corolla throat.

M. layneae

M. breweri (Greene) Cov. Brewer's Monkey-flower. Glandular-pubescent annual; stems simple, 3–10 cm tall; lvs. narrowly elliptic to oblong, entire, 3–20 mm long; pedicels 3–10 mm long; calyx 5–7 mm long, ± scarious between the 5 glandular-pubescent ridges, the teeth ca. 0.5 mm long, deltoid; corolla pale pink to purplish, 6–10 mm long, persistent and shriveling on the capsule. Found at one location. Open, damp meadow, 5,000 ft. Mixed Conifer F. Big Flat (*511*). June.

M. guttatus Fisch. ex DC. Common Large Monkey-flower. Annuals or perennials, often rooting at the nodes and forming stolons; stems glabrous below the infl., stout to weak, simple to branched, tending to be hollow in larger plants, 1–5 dm tall; lvs. oval to round, denticulate, 1–4 cm long, the lower petioled, the upper becoming sessile and ± connate; infl. racemose, the pedicels 1–6 cm long; calyx 1–2 cm long, often spotted or tinged with red, becoming inflated in age, the lower teeth curving upward, the uppermost tooth ca. 2 times longer than the others; corolla 1.5–4 cm long, yellow with red spots, the throat nearly closed by 2 hairy ventral ridges. Common. Damp or wet, usually shaded places, 5,000–7,300 ft. Mixed Conifer F., Red Fir F., Subalpine F. Big Flat (*225*), Caribou Mt. (*401*), Caribou Basin (*1075*), Grizzly L. Basin (*894*), head of Union Cr. (*1001*). July–August.

ssp. *micranthus* (Heller) Munz. Delicate annuals, 4–10 cm high; pedicels 1–3 cm long; calyx 5–10 mm long, the lower teeth only slightly curved to straight; corolla 10–15 mm long. Occasional. Damp meadows, 5,000 ft. Mixed Conifer F. Big Flat (*97, 506*). June.

M. layneae (Greene) Jeps. Layne's Monkey-flower. Slender, glandular-pubescent annual; stems simple, 3–10 cm tall; lvs. oblong to linear, ca. 1 cm long, finely glandular; pedicels 1–2 mm long; calyx 5–6 mm long, ± inflated and ridged in age, the teeth and ridges glandular-pubescent; corolla red-purple with pale ventral ridges in throat, 13–20 mm long. Found at one location. Dry, sandy flats, 5,600 ft. Montane Chaparral. Canyon Cr. (*767*). August.

M. lewisii Pursh. Lewis' Monkey-flower. Fig. 76. Stout perennial from trailing, matted rootstocks; stems erect, simple to ± branched, 3–8 dm tall, glandular-pubescent; lvs. oblong-elliptic, sinuate-denticulate, 2–7 cm long, mostly sessile with round ± clasping bases, glandular; pedicels 3–9 cm long, from the upper axils; calyx 15–25 mm long with triangular-subulate teeth 4–6 mm long; corolla 2.8–4 cm long, rose to pink with 2 yellow, hairy, ventral ridges in the throat. Occasional. Rocky stream banks, 6,200–7,800 ft. Red Fir F., Subalpine F., Alpine Fell-fields. Caribou Basin (*386*), Boulder Cr. Basin (*477*), Canyon Cr. (*810*), north side of Thompson Pk. (*905*). August.

M. moschatus Dougl. ex Lindl. Musk Flower. Perennial from slender, creeping rootstocks; stems 0.5–3 dm long, simple to diffusely branched, densely pilose, ± slimy; lvs. ovate to somewhat oblong, 1–4 cm long, denticulate, petioled or the upper becoming sessile; pedicels slender, pilose, 1.5 cm long; calyx 9–12 mm long, pilose, the teeth lanceolate, 2–3 mm long; corolla yellow with red lines and spots in throat, 18–28 mm long, the tube exserted from the calyx. Common. Damp, rocky or sandy, shaded places, 5,000–6,000 ft. Mixed Conifer F., Red Fir F. Caribou Mt. (*216*), Josephine L. (*282*), head of South Fork of Salmon R. (*410*), Yellow Rose Mine (*673*), Canyon Cr. (*793*), Big Flat (*1035*). July–August.

var. *moniliformis* (Greene) Munz. Differing from the above in being glabrous to finely pubescent below the infl. and all of the lvs. having petioles. Occasional. Damp, shaded places, 6,500 ft. Red Fir F. Caribou Basin (*736*), Kidd Cr. (*1104*). July–August.

M. nanus H. & A. Dwarf Purple Monkey-flower. Glandular-pubescent annuals; stems simple to branched, 3–15 cm tall, densely glandular; lvs. narrowly oblanceolate, short petioled to sessile, 1–2 cm long; pedicels ca. 2 mm long; fls. numerous on larger plants; calyx 5–7 mm long, glandular-pubescent on the lobes and ridges, scarious below the sinuses, the lobes ca. 2 mm long; corolla purple, often pale in the throat, 12–17 mm long. Occasional. Dry, open slopes and flats, 5,000–5,500 ft. Mixed Conifer F. Yellow Rose Mine Trail (*137*), head of South Fork of Salmon R. (*409*), Packers Pk. (*552*). June–August.

Boykinia major

Parnassia palustris var. californica

75

Ribes roezlii

Mimulus lewisii

M. primuloides Benth. Primrose Monkey-flower. Short stemmed, rhizomatous perennial; stems 1–5 cm long, villous to subglabrous; lvs. crowded, oblong to obovate, sharply toothed to entire, ± cuneate at the base, 1–4 cm long; pedicels slender, erect, 3–8 cm long; calyx 6–8 mm long, the teeth apiculate, 0.5–1 mm long; corolla yellow with red spots, 15–20 mm long. Occasional. Wet stream banks and meadows, 6,500–7,000 ft. Red Fir F. Landers L. (*986*), head of Union Cr. (*1002*). August.

ssp. *linearifolius* (Grant) Munz. Lvs. oblanceolate, 1–4 cm long; pedicels 6–8 cm long; calyx 7–8 mm long; corolla 15–20 mm long. Found at one location. Bog, 6,500 ft. Red Fir F. Yellow Rose Mine (*955*). August.

M. pulsiferae Gray. Candelabrum Monkey-flower. Small, delicate annuals; stems 3–10 cm tall, glandular-puberulent, simple to loosely branched; lvs. short petioled, 4–10 mm long; pedicels 1–2 cm long, slender, ± curved and spreading in fruit; calyx 4–6 mm long with acute lobes; corolla yellow with red spots, 8–10 mm long. Found at one location. Open meadow, 5,000 ft. Mixed Conifer F. Big Flat (*586*). June.

M. tilingii Regel. Larger Mountain Monkey-flower. Perennial from a mass of slender, yellowish rootstocks; stems decumbent to erect, 1–2 dm long, glabrous to sparsely pilose; lvs. slimy, ± glandular-puberulent, 1–3 cm long, sinuately dentate, the lower short petioled, the upper sessile; pedicels stout, 3–4 cm long, glandular-pilose; calyx reddish, 1–1.5 cm long, becoming inflated in fruit, the uppermost tooth ca. 2 times longer than the others; corolla yellow with red spots, 2.5–3 cm long. Common. Gravelly, wet soil, 6,200–7,800 ft. Subalpine F., Alpine Fell-fields. North side of Thompson Pk. (*918*), Caribou Basin (*756, 1056*). August. This species has been previously reported in Calif. only from the Sierra Nevada and the southern mountains (Munz, 1959).

Orthocarpus Nutt.

O. copelandii Eastw. Copeland's Orthocarpus. Erect annuals; stems simple to branched, 1–4 dm tall, finely canescent; lvs. linear-acuminate, 1–4 cm long, entire or the upper with linear lobes, sessile; spikes 2–7 cm long, densely fld.; bracts abruptly different from lvs., ca. 1 cm long with a nearly basal pair of attenuate lobes, the mid-lobe 5–8 mm wide, lavender tipped, scabrid on the margins; calyx 4–5 mm long with lanceolate lobes; corolla 10–15 mm long, rose-purple with white lower lip, galea pubescent. Common. Mixed Conifer F., Red Fir F., Subalpine F. Big Flat (*219, 238, 707*), Yellow Rose Mine (*958*), Red Rock Mt. (*1021*). July–August.

Pedicularis L.

Calyx lobes seemingly 2 . . . . *P. racemosa*
Calyx lobes 5.
Tip of galea blunt and beakless.

Corolla greenish white to pale yellow. . . . . *P. flavida*
Corolla reddish orange . . . . *P. densiflora*
Tip of galea extended into a slender, curving beak.
Corolla pale yellow with purple spots in throat. . . . *P. contorta*
Corolla pale lavender with dark purple lines. . . . . . *P. attolens*

P. attolens Gray. Little Elephant's Head. Scapose perennial; stems glabrous below infl., 1.5–4 dm tall; lvs. basal, the blades deeply pinnatifid, 3–9 cm long, the petioles shorter; bracts becoming reduced up the stem; infl. narrow, 4–10 cm long, woolly pubescent; calyx ca. 5 mm long with 5 lance-linear lobes; corolla pale purple with dark purple lines, the tube 5–6 mm long, the galea extended into a narrow upcurved beak longer than the body. Common. Damp meadows, 6,500–7,200 ft. Red Fir F. Caribou Basin (*396*, *1063*), ridge west of Bullards Basin (*965*), Ward L. (*1090*). August.

P. contorta Benth. ex Hook. White Coiled-beak Lousewort. Glabrous perennial; stems 2–4 dm tall; lvs. basal and on the stem, 3–9 cm long, pinnately divided, the segments sharply toothed, stem lvs. reduced upwards; infl. 2–12 cm long, bracts narrow, few lobed; calyx 6–8 mm long with 5 lanceolate lobes; corolla cream to yellowish with purple spots, ca. 15 mm long, the galea extended into a decurved beak shorter than the body. Common. Damp, open meadows, 6,000–7,500 ft. Red Fir F., Subalpine F. Caribou Basin (*397*, *740*, *1064*), Canyon Cr. (*802*), Grizzly L. Basin (*909*). July–August.

P. densiflora Benth. ex Hook. Indian Warrior. Perennial from a woody rootstock; stems ca. 1 dm tall, pubescent, exceeding lvs.; lvs. mostly basal, 4–8 cm long, deeply pinnately divided, the segments sharply toothed; bracts ca. as long as fls., sharply toothed toward the tip; infl. dense, 3–6 cm long; calyx deep red, 8–12 mm long with 5 pubescent, lance-ovate lobes; corolla reddish orange, 2–2.5 cm long, the lower lip much shorter than the bluntly tipped galea. Occasional. Dense woods, 5,000 ft. Mixed Conifer F. Big Flat (*596*). June.

P. flavida Penn. Cascade Mountains Lousewort. Glabrous perennial; stems 3–6 dm tall; lvs. 5–10 cm long, deeply lobed into sharply toothed segments, the lower segments distinct, the upper becoming confluent; infl. rather dense, 3–8 cm long, bracts lanceolate, shorter than the fls., entire to dentate; calyx ca. 6 mm long with 5 lanceolate lobes; corolla greenish white to pale yellow, 15–20 mm long, the galea strongly hooded, blunt at apex, the lower lip ca. half as long as galea. Common. Damp slopes and meadows, 6,000–7,400 ft. Red Fir F. Caribou Basin (*387*, *725*), Canyon Cr. (*797*), Grizzly L. Basin (*883*). July–August.

P. racemosa Dougl. ex Hook. Leafy Lousewort. Glabrous, branched perennial with several stems from a ± woody rootcrown; stems 3–5 dm tall with slender branches; lvs. lanceolate, serrate, 3–5 cm long on short petioles; calyx ca. 5 mm long with 2 broad lobes; corolla whitish to purple tinged, 10–12 mm long, the galea extended into a slender, hooked beak, the lower lip deflexed-spreading, ca. 10 mm wide at the tip. Found at one location. Dry, shaded slope in woods, 6,200–6,500 ft. Red Fir F. Kidd Cr. (*443, 1101*). August.

Penstemon Mitch.

Plants ± shrubby, 5–15 dm tall . . . . *P. lemmonii*
Plants not shrubby or, if woody at the base, then much shorter.
  Anthers densely comose.
    Plants low, decumbent, matted; corolla glabrous externally.
      Corolla rose-purple . . . . *P. newberryi* ssp. *berryi*
      Corolla blue-violet . . . . *P. davidsonii*
    Plants taller, erect; corolla glandular-pubescent externally. *P. nemorosus*
  Anthers glabrous to puberulent.
    Anther sacs opening across their continuous apices, the free tips remaining saccate and ± parallel after dehiscence.
      Lvs. glabrous to sparingly pubescent. *P. laetus* ssp. *roezlii*
      Lvs. densely canescent. . . . . *P. purpusii*
    Anther sacs opening from their free tips throughout, ± widely spreading after dehiscence.
      Corolla pinkish to rose . . . . *P. tracyi*
      Corolla blue to purple or buff.
        Corolla pale yellow to buff . . . . *P. deustus*
        Corolla blue to purple.
          Infl. glandular-pubescent; corolla 13–18 mm long . . . . *P. anguineus*
          Infl. not glandular-pubescent; corolla 8–11 mm long . . . . *P. procerus* ssp. *brachyanthus*

P. anguineus Eastw. Siskiyou Penstemon. Perennial herbs with spreading stems; the stems glabrous below infl., 2–5 dm tall; lvs. ovate, 1–4 cm long, the lower petioled, the upper sessile, finely denticulate to subentire; infl. glandular-pubescent, of several dense clusters of fls. or more openly paniculate; calyx 4–6 mm long with lanceolate lobes; corolla blue-purple, 13–18 mm long, the upper lip short and erect, the lower lip longer and spreading; anther sacs puberulent, ± divaricate after dehiscence; staminode exserted, sparsely bearded to glabrous. Occasional. Dry woods, 5,000–6,500 ft. Mixed Conifer F., Red Fir F. Caribou Mt. (*325*), Big Flat (*703*). July–August.

P. davidsonii Greene. Creeping Penstemon. Low perennial, forming creeping mats from a branched, woody caudex; flowering stems mostly less than 1 dm high, glabrous below becoming puberulent upwards; lvs. elliptic to round, 5–10 mm long, thickish, ± glandular-punctate, with narrow petioles 1–5 mm long; infl. few-fld., glandular-pubescent; calyx 7–11 mm long, the lobes linear to lanceolate; corolla blue-purplish, 18–30 mm long; anthers densely comose, staminode bearded, half as long as fertile stamens. Occasional. Shaded, rocky ridges, 7,800–9,000 ft. Subalpine F. Thompson Pk. (*822*), Packers Pk. (*1393*). July–August.

P. deustus Dougl. ex Lindl. Hot-rock Penstemon. Perennial with woody stems branched below; stems 2–5 dm tall, erect, glabrous to glandular-puberulent; lvs. elliptic to ovate, 1–3 cm long, coarsely serrate, the basal crowded and petioled, the cauline sessile, in 5–12 pairs; infl. narrow, of several dense clusters, many fld.; calyx 4–6 mm long, the lobes lanceolate to ovate; corolla pale yellow to buff with purple lines, 10–16 mm long, spreading to ascending, sparingly glandular; anthers glabrous, ± divergent after dehiscence; staminode usually glabrous, sometimes short bearded. Common. Dry, rocky slopes, 5,500–6,000 ft. Mixed Conifer F., Red Fir F. Boulder Cr. Basin (*175*), Kidd Cr. (*678*). July.

P. laetus Gray ssp. *roezlii* (Regel) Keck. Gay Penstemon. Perennial from a woody base; stems 1–4 dm tall, erect to spreading, finely pubescent to puberulent below the infl.; lvs. crowded at base, becoming fewer up the stem, linear to lanceolate, 2–10 mm wide, often folded lengthwise; infl. glandular-pubescent, narrow to somewhat spreading; calyx 4–7 mm long, the lobes lanceolate to narrowly ovate; corolla blue-purple, ca. 15 mm long, sparsely glandular without; anthers glabrous, coarsely toothed along the suture, remaining saccate; staminode glabrous. Occasional. Dry, rocky slopes and meadows, 6,500–7,000 ft. Red Fir F. Yellow Rose Mine (*299*), ridge west of Bullards Basin (*962*), head of Union Cr. (*989*). August.

P. lemmonii Gray. Bush Beard-tongue. Open, ± shrubby plants 5–15 dm tall; the stems brown, glabrous; lvs. elliptic to lance-ovate, 1–6 cm long, 0.5–2.5 cm wide, serrulate, short petioled; infl. narrow, the fls. borne in several openly spaced clusters; pedicels densely glandular-pubescent; calyx 4–7 mm long with lanceolate lobes, densely glandular-pubescent; corolla 10–12 mm long, yellowish brown with a dark brown upper lip that is broadly galeate; anthers glabrous, divergent; the staminode densely yellow bearded and exserted. Occasional. Rocky cliff, 5,000–5,500 ft. Mixed Conifer F. Above McKay Camp, Canyon Cr. (*860*). August.

P. nemorosus (Dougl. ex Lindl.) Trautv. Woodland Penstemon. Stems erect, 3–4 dm tall, puberulent at least below; lvs. all cauline, equally spaced, lanceolate to ovate with rounded bases, 2–5 cm long, short petioled, serrate; infl. few fld., pedicels glandular-pubescent; calyx ± glandular-pubescent, 6–10 mm long, the lobes lanceolate to ovate; corolla purple to rose, 2–3 cm long, glandular-pubescent, the lower lip longer than the upper; anthers densely tawny comose, the filaments retorsely puberulent above, hirsute below; staminode ca. three-quarters as long as fertile filaments, bearded. Found at one location. Wooded, rocky slope, 7,500 ft. Red Fir F. Sawtooth Ridge (*1076*). August.

P. newberryi Gray ssp. *berryi* (Eastw.) Keck. Mountain Pride. Fig. 77. Low perennial from a woody, spreading, matted base; stems 1–3 dm tall, glabrous to puberulent; lvs. coriaceous, serrulate to serrate, elliptic to orbicular, 5–20 mm long, short petioled becoming reduced and sessile on flowering stem; infl. few to several fld., glandular-pubescent; calyx 7–12 mm long, the lobes lanceolate, ± glandular-pubescent; corolla rose-purple, 27–33 mm long, white bearded on lower lip and ventral throat; anthers densely comose, included; staminode one-half to three-quarters as long as fertile filaments, mostly bearded. Common. Open, dry, rocky slopes, 5,600–8,000 ft. Mixed Conifer F., Red Fir F., Subalpine F. Josephine L. (*75*), Canyon Cr. (*170*), Caribou Basin (*194, 752*), Caribou Mt. (*333*), Boulder Cr. Basin (*478*), Kidd Cr. (*679, 1100*), ridge south of Thompson Pk. (*819*), Sawtooth Ridge (*1068*). June–August.

P. procerus Dougl. ex Grah. ssp. *brachyanthus* (Penn.) Keck. Small-flowered Penstemon. Perennial, forming dense clumps from a branched caudex; stems erect, 1.5–3 dm tall, glabrous; lower lvs. 1.5 cm long, ovate to oblanceolate, often folded, on long, narrow petioles, cauline lvs. sessile, few to several pairs; infl. consisting of 1–several dense clusters; calyx 2–3 mm long, the lobes with prominent scarious, erosulate margins; corolla blue-purple, 8–11 mm long, horizontal to somewhat deflexed; anthers glabrous, ± divaricate, staminode bearded with few to many stout hairs. Common. Open, often rocky slopes and meadows, 5,000–7,200 ft. Mixed Conifer F., Red Fir F. Josephine L. (*81*), Canyon Cr. (*162, 858*), Caribou Basin (*195, 398*), Kidd Cr. (*681*), Grizzly L. Basin (*889*), Ward L. (*1091*). June–August.

P. purpusii Bdg. Purpus' Penstemon. Perennial from a ± woody rootcrown; stems 1–2 dm long, erect to decumbent, canescent; lvs. obovate to narrowly oblanceolate, often folded, densely canescent, 1–3.5 cm long; infl. glandular-pubescent; calyx 5–10 mm long, glandular-pubescent; corolla blue-purple, 20–30 cm long, sparsely glandular-pubescent; anther sacs opening ca. three-quarters of their length, remaining saccate, glabrous except for tuft of hairs at sinus, spinose on suture margin; staminode ca. as long as fertile filaments, glabrous. Occasional. Dry, open slopes, 6,000–6,500 ft. Red Fir F. Packers Pk. (*557*), Yellow Rose Mine (*672*). June–July.

P. tracyi Keck. Tracy's Penstemon. Fig. 78. Perennial from a woody decumbent base; stems 10–15 cm tall, pubescent; lvs. cuneate-oblong to oval or round, 1–2 cm long, entire to finely denticulate, the basal petioled, the cauline becoming sessile; infl. dense and contracted; calyx 2–3 mm long, the lobes with narrowly hyaline, erosulate margins; corolla pinkish, darker on the upper lip, 10–14 mm long, ± deflexed, glabrous except for the densely bearded lower lip; anthers glabrous, divaricate; staminode sparsely bearded, ca. two-thirds as long as fertile filaments. Rare. Dry, rocky ledge, 6,800 ft. Red Fir F. Packers Pk. (*550, 937*). June–August. This species is apparently a local endemic, previously reported only from the Devil's Canyon Mts., the type locality, ca. 15 miles to the west (Keck, 1940).

Veronica L.

Main stem ending in a single racemelike infl.
  Corolla 10–12 mm broad; style ca. 9 mm long. . . . . . . *V. copelandii*
  Corolla 5–7 mm broad; style shorter.
    Lvs. densely pubescent. . . . . *V. alpina* var. *alternifolia*
    Lvs. glabrous to sparingly pubescent.
      V. serpyllifolia var. *humifusa*
Main stem with lateral racemes below the tip. . . . . *V. americana*

V. alpina L. var. *alternifolia* Fern. American Alpine Speedwell. Stems slender, erect, loosely pilose, 1–2 dm tall; lvs. 1–2 cm long, ovate to elliptic, densely pubescent, entire; infl. elongating in fruit; pedicels 2–4 mm long; calyx ca. 3 mm long, pilose; corolla blue, 5–7 mm wide; capsule retuse at tip, pilose. Occasional. Wet meadows and stream banks, 6,000–8,200 ft. Red Fir F., Subalpine F. Boulder Cr. Basin (*476*), north side of Thompson Pk. (*1306*). August.

V. americana (Raf.) Schw. American Brooklime. Glabrous ± succulent perennial from creeping, horizontal stems which root at the nodes; flowering stems decumbent to erect, 1–2 dm long; lvs. lance-ovate to lanceolate, serrate to denticulate, 1.5–3 cm long; racemes lax, few to many fld., lower pedicels slender, up to 12 mm long; calyx 2–3 mm long; corolla rose with darker lines, 5–6 mm broad; capsule rounded, 3–4 mm long, glabrous. Common. Wet, boggy meadows and seeps, 5,000–6,500 ft. Mixed Conifer F., Red Fir F. Yellow Rose Mine (*132*), Big Flat (*489, 527*). June.

V. copelandii Eastw. Copeland's Speedwell. Fig. 79. Perennial forming dense clumps; stems erect, 1–2 dm tall, glandular-villous, slimy to the touch; lvs. sessile, in ca. 5 pairs, densely glandular-villous, lance-ovate, 1–3 cm long; racemes up to 12 cm long, glandular-villous; pedicels ca. 1 cm long in fruit; calyx 4 lobed, 3–4 mm long, glandular-pubescent; corolla deep blue, 10–12 mm broad; stamens 2 or 3; capsule 4–5 mm long, glandular, notched at apex; style ca. 9 mm long. Rare. Rocky serpentine slopes and meadows, 7,500 ft. Subalpine F. Red Rock Mt. (*1020, 1032*). July–August.

V. serpyllifolia L. var. *humifusa* (Dickson) Vahl. Thyme-leaved Speedwell. Stems decumbent to erect, 1–2 dm tall, strigulose becoming glandular above; lvs. 5–15 mm long, elliptic to ovate, glabrous to sparingly pubescent, becoming sessile above, crenulate; infl. elongating in fruit; pedicels 3–5 mm long;

calyx 3–4 mm long; corolla blue, 5–7 mm broad; capsule broader than long. Common. Wet meadows and seeps, 5,000–6,200 ft. Mixed Conifer F., Red Fir F. Big Flat (*523, 488*), Yellow Rose Mine (*133*). June.

SOLANACEAE

Solanum L.

S. *parishii* Heller. Parish's Nightshade. Low, spreading, shrubby perennial; stems 3–4 dm tall, angled, glabrous; lvs. lanceolate to oblanceolate, glabrous, 1–2 cm long, tapering to short petioles; fls. many on slender petioles, 1.5–2 cm long; calyx ca. 4 mm long, the deltoid lobes ca. equal to the tube; corolla blue, rotate, shallowly lobed, ca. 2 cm broad; anthers 4–5 mm long; berries 7–9 mm in diam. Occasional. Dry rocky slopes and woods, 5,000–5,800 ft. Mixed Conifer F. Kidd Cr. (*680*). July.

UMBELLIFERAE

Fr. bearing tubercles or spines .... *Sanicula*
Fr. not bearing tubercles or spines.
  At least some of the ribs of the fr. winged at maturity.
    Lateral ribs winged, dorsal ribs filiform.
      Plants to 20 dm tall...................... *Heracleum*
      Plants 1–4 dm tall .... *Lomatium*
    Lateral and dorsal ribs winged.
      Wings prominent .... *Angelica*
      Wings narrow, inconspicuous .... *Ligusticum*
  None of the ribs winged.
    Fr. elongate, 10–15 mm long .... *Osmorhiza*
    Fr. subglobose, 2–3 mm long.
      Plants with stems above ground, 4–6 dm high. *Perideridia*
      Plants with most of stem below ground, 0.5–1 dm high. *Orogenia*

Angelica L.

A. arguta Nutt. ex T. & G. Lyall's Angelica. Stout perennial to 1.5 m tall; stems ± spreading from the base; lvs. 10–20 cm long, ovate in outline, twice pinnate, the lfts. ovate to lanceolate, coarsely serrate, 3–8 cm long, 1–3.5 cm wide, ± stiff-pubescent on the midribs, otherwise glabrous; invol. none; involucel absent or of few–several narrow bractlets 2–3 mm long; rays 25–40, subequal, 3–10 cm long; pedicels 2–12 mm long, webbed at base; petals white, ± scabrous; ovaries ± scabrous; fr. ca. 8 mm long, the dorsal ribs narrowly winged, the lateral ribs broadly winged. Common. Dry slopes and meadows, 5,000–8,000 ft. Montane Chaparral, Mixed Conifer F., Red Fir F., Subalpine F. Big Flat (*244*), Yellow Rose Mine (*945*), Red Rock Mt. (*1026*).

77

Penstemon newberryi ssp. berryi

78

Penstemon tracyi

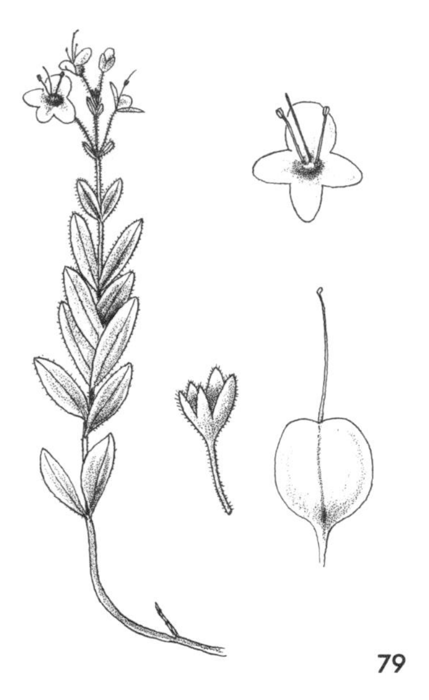

79

Veronica copelandii

80

Ligusticum californicum

August. The scabrous petals and ovaries of this material resemble some forms of *A. tomentosa*, yet the shape of the lfts. and webbed pedicels are typical of *A. arguta*.

Heracleum L.

H. lanatum Michx. Cow-parsnip. Stout perennial 1–2 m tall; lf.-blades round to reniform in outline, 2–5 dm long, the lfts. ovate to roundish, 1.5–2 dm long, ± lobed and coarsely serrate; invol. bracts 5–15 mm long; involucel bractlets lance-acuminate, 5–15 mm long; rays 5–10 cm long; pedicels 1–1.5 cm long; fls. white; fr. glabrous, obovate to obcordate, ca. 1 cm long, the lateral ribs winged, dorsal ribs filiform, oil tubes prominent, extending ca. halfway to base. Occasional. Damp meadows, 5,000–5,700 ft. Mixed Conifer F. Canyon Cr. (*792*). July–August.

Ligusticum L.

L. californicum Coult. & Rose. California Lovage. Fig. 80. Glabrous perennial from a stout, fibrous root crown with an odor of celery; stems 4–7 dm tall, simple and naked below, branched above; lvs. bipinnate, the lfts. ovate to obovate, 1–2 cm long, ± divided into linear lobes; rays 3–5 cm long; pedicels slender, 5–10 mm long; fls. yellowish white; fr. oval, 4–5 mm long, the ribs narrowly winged. Common. Open, rocky meadows and slopes, 5,000–7,000 ft. Mixed Conifer F., Red Fir F. Ward L. (*449*), Canyon Cr. (*796, 851*), Landers L. (*983*), head of Kidd Cr. (*1098*). August.

Lomatium Raf.

L. macrocarpum (H. & A.) Coult. & Rose. Large-fruited Lomatium. Acaulescent perennial from a deep, slender taproot; lvs. basal, 4–9 cm long, pubescent, ternate, the main lfts. then pinnately divided, the ultimate divisions linear to oblanceolate, 1–4 mm long; petioles 3–5 cm long; fruiting peduncles 1.5–2.5 dm high, purplish, pubescent; primary rays of umbel 1–6 cm long; fr. 10–15 mm long at maturity with prominent lateral wings. Common. Dry, rocky places, 5,500–8,000 ft. Mixed Conifer F., Red Fir F., Subalpine F. Packers Pk. (*1397*), Black Mtn. (*1401*), ridge between Landers L. and Union L. (*1470*). July.

Orogenia Wats.

O. fusiformis Wats. California Orogenia. Glabrous perennial from a fusiform taproot; plants 5–10 cm high, stems mainly underground; lvs. ternately divided into linear segments 1–6 cm long; fls. white in small, compound umbels borne on peduncles 5–10 cm long; fr. oval, 3 mm long. Found at one location. Open woods, 5,000 ft. Mixed Conifer F. Big Flat (*1128*). June.

Osmorhiza Raf.

Fr. densely hispid at the attenuate base, tapering toward a narrow beak at apex . . . . *O. chilensis*

Fr. glabrous, not attenuate at base, wider at apex. . . . . . . *O. occidentalis*

O. chilensis H. & A. Mountain Sweet-cicely. Slender perennial from a thickened root; stems 3–6 dm tall, canescent below becoming less so above; lvs. rounded in outline, biternate, the lfts. ovate, 1.5–3 cm long, coarsely toothed to lobed; invol. and involucel reduced to wanting; rays few, slender, 3–9 cm long; pedicels 5–15 mm long; fls. greenish white to white; fr. 10–15 mm long, densely hispid at the narrow base, tapering toward the beak at the apex. Common. Dry woods, 5,000–6,000 ft. Mixed Conifer F. Big Flat (*704*), Grizzly Cr. (*874*). July–August.

O. occidentalis (Nutt.) Torr. Western Sweet-cicely. Slender stemmed perennial 5–6 dm high; lvs. 4–12 cm long, once or twice divided into lanceolate to ovate, coarsely toothed lfts. 1.5–5 cm long with a licorice odor when crushed; fls. yellow; fr. ca. 15–17 mm long, glabrous, only slightly tapered at the ends. Occasional. Cool, shaded, rocky places, 5,500–7,300 ft. Mixed Conifer F., Red Fir F., Subalpine F. Landers L. Basin (*1480*), Canyon Cr. (*Alexander & Kellogg 5458*). July.

Perideridia Rchb.

P. parishii (Coult. & Rose) Nels. & Macbr. ssp. *latifolia* (A. Gray) Chuang & Constance. Parish's Yampah. Glabrous perennial from a solitary, fusiform tuber or small cluster of tubers; stems slender, 4–6 dm tall; lvs. pinnate into linear to lanceolate entire divisions 2–12 cm long; bracts of invol. few or absent; bractlets of involucel several, linear, 1–3 mm long; rays many, 1–2 cm long; pedicels 3–5 mm long; fls. white; fr. subglobose, 2–3 mm long, the ribs filiform. Common. Wet meadows and bogs, 5,000–7,000 ft. Mixed Conifer F., Red Fir F. Big Flat (*233, 371*), Yellow Rose Mine (*946*). August.

Sanicula L.

Fr. bearing tubercles armed with hooked spines ....S. *graveolens*
Fr. bearing unarmed tubercles..........................S. *tuberosa*

S. *graveolens* Poepp. ex DC. Sierra Sanicle. Perennial from a fusiform taproot; stems branched above, purplish, 15–25 cm high; lvs. mostly on lower stem, ternate, 1.5–4 cm long; bracts 2, pinnately divided; rays 3–4, unequal, 1–4 cm long; fr. ovoid to globose 3–4 mm long, the swollen tubercles armed with hooked bristles. Occasional. Dry slopes, 5,000–8,000 ft. Mixed Conifer F., Montane Chaparral. Packers Pk. (*1149, 1396*). June–July.

S. *tuberosa* Torr. Tuberous Sanicle. Perennial from a small, deep-seated tuber; stems divided near base, ascending, 5–15 cm high; lvs. 3–5 cm long, tripinnate, the ultimate divisions linear, cauline lvs. reduced upwards; bracts 2, mostly pinnatifid; bractlets several, linear, 1–3 mm long; rays mostly 3, unequal, 7–15 mm long; pedicels 1–5 mm long; fls. yellow; fr. ovoid to subglobose, 1.5–2 mm long, bearing inflated, unarmed tubercles. Occasional. Dry, rocky meadows, 5,000 ft. Mixed Conifer F. Along the South Fork of the Salmon R. (*589*). June–July.

VALERIANACEAE

Corolla pink; plants annual, 1–2 dm high. . . . . *Plectritis*
Corolla greenish-white; plants perennial, 2.5–4 dm high. . . . . . *Valeriana*

Plectritis DC.

P. congesta (Lindl.) DC. ssp. *brachystemon* (F. & M.) Morey. Pink Plectritis. Slender stemmed annual, 10–15 cm high; lvs. opposite, oblong-elliptic, 1–2 cm long, sessile; fls. numerous, borne in a dense, bracteate head; corolla pink, 2–4 mm long, with a small spur at the base of the tube; fr. a winged achene, 2.5–3 mm long, tan, keeled on the side opposite the wings which are connivent at base, divergent at the summit and pubescent with stiff hairs. Found at one location. Damp bank, 5,000 ft. Mixed Conifer F. Big Flat (*1138*). June.

Valeriana L.

V. capitata Pall. ex Link. ssp. *californica* (Heller) F. G. Mey. California Valerian. Perennial from a slender, creeping rootstock; stems 3–4 dm high; basal lvs. loosely tufted, spatulate to elliptic, the blades entire, 2–5 cm long on long, slender petioles; cauline lvs. nearly sessile, 4–7 cm long, deeply divided into toothed, oblanceolate segments; corolla funnelform, greenish-white, pubescent within, 3 mm long, slightly gibbous at base; achenes ovate, purplish, 5–6 mm long, crowned with ca. 15 plumose bristles. Found at one location. Open, rocky places, 7,000–7,500 ft. Red Fir F. Landers L. Basin (*1481*). July.

VERBENACEAE

Verbena L.

V. lasiostachys Link. Western Verbena. Much branched perennial; stems hirsute, 5–8 dm high; lvs. ovate, 4–8 cm long, toothed to moderately incised, the petioles winged; spikes borne at the ends of branches becoming 15–25 cm long in fr., quite hairy and glandular; calyx 3–4 mm long subtended by a lanceolate bract 3 mm long; corolla purple, the tube 4–5 mm long; nutlets 1.5 mm long, reticulate on backs. Occasional. Dry, open, rocky areas, 5,000 ft. Mixed Conifer F. Big Flat (*1399*). July–August.

VIOLACEAE

Viola L.

Corollas yellow.
    Lvs. palmately divided into linear segments. . . . . *V. sheltonii*
    Lvs. crenately toothed to entire.
        Upper petals purplish brown on back. . . . . *V. purpurea*
        Upper petals yellow on back. . . . . *V. glabella*
Corollas white to blue or purple.
    Corollas blue . . . . *V. adunca*
    Corollas white to lavender.
        Upper petals purplish red on back. . . . . *V. cuneata*

Upper petals white on back. . . . . *V. macloskeyi*

V. adunca Sm. Western Dog Violet. Puberulent perennial from slender rootstocks; stems 3–5 cm long; lvs. ovate to round-ovate, ± cordate and crenulate, 1–2.5 cm long, petioles slender, 2–4 cm long, stipules linear-lanceolate, lacerate toothed; peduncles shorter to longer than petioles; sepals linear-lanceolate, ca. 5 mm long; corollas blue with purple veins, 8–13 mm long, the spur purple, ca. 5 mm long, the lateral petals white-bearded toward base; capsule 6–8 mm long. Occasional. Open to shaded meadows, 5,000 ft. Mixed Conifer F. Along South Fork of the Salmon R. (*98, 588*). June–July.

V. cuneata Wats. Wedge-leaved Violet. Glabrous perennial from thickened rootstocks; stems slender, 3–6 cm long; lvs. round-ovate, crenate, 1.5–2.5 cm long, petioles 3–5 cm long, stipules ± entire; peduncles scarcely exceeding lvs.; sepals lanceolate, ca. 3 mm long; petals 6–8 mm long, white to pale lavender with purple veins, the upper and lateral pairs with a purple spot near the base, spur yellow, all petals deep purple-red on back; capsule subglobose, ca. 5 mm in diam. Found at one location. Damp, rocky meadow, 6,800 ft. Subalpine F. Landers L. (*661*). July.

V. glabella Nutt. Stream Violet. Glabrous perennial from a thickened rootstock; stems slender, 5–15 cm tall; basal lvs. cordate, crenate, on petioles 4–15 cm long; cauline lvs. usually smaller on short petioles borne only at the summit of the stem; peduncles little exceeding cauline lvs.; sepals lanceolate, 5–6 mm long; petals yellow with purple veins, 6–16 mm long, spur 1–2 mm long; capsule 7–8 mm long. Common. Damp shaded places, 5,000–6,500 ft. Mixed Conifer F., Red Fir F. Along South Fork of the Salmon R. (*104*), Canyon Cr. (*155*), Big Flat (*533*), Union L. Basin (*642*), head of Sunrise Cr. (*664*). June–July.

V. macloskeyi Lloyd. Macloskey's Violet. Fig. 81. Low, subglabrous, acaulescent perennial from slender, creeping rootstocks; lvs. round-reniform, entire to crenulate, 1–2 cm long, petioles 2–3 cm long; peduncles 3–6 cm long; sepals ovate-lanceolate, 3–4 mm long; petals 6–9 mm long, white, the lower 3 with purple veins, spur 2–3 mm long; capsule ca. 6 mm long. Common. Wet meadows, 5,000–6,500 ft. Mixed Conifer F., Red Fir F. Canyon Cr. (*61*), Yellow Rose Mine (*131*), Bullards Basin (*633*). June–July.

V. purpurea Kell., not Stev. Mountain Violet. Perennial from a woody taproot; stems 3–6 cm tall, retrorse-pubescent; lvs. ovate, crenately toothed, cordate to cuneate at base, 1–2 cm long, pubescent, green above, purplish gray beneath, petioles 3–6 cm long; peduncles 3–7 cm long; petals yellow, 8–10 mm long, the lower 3 with purple veins, the upper with purple on the back; capsule 5–6 mm long. Common. Dry, rocky slopes, 5,000–7,000 ft. Mixed Conifer F., Red Fir F. Big Flat (*496*), Packers Pk. (*548*), Landers L. (*657*). June–July.

V. sheltonii Torr. Shelton's Violet. Glabrous perennial from a short rootstock; stems very short; lvs. rounded in outline, 1–3 cm broad, palmately divided into linear segments, petioles slender, 3–5 cm long; peduncles 4–8 cm long; sepals lanceolate, 5–6 mm long; petals yellow, ca. 10 mm long, the 3 lower with purple veins, the upper with purple-brown backs; capsule 7–8 mm long. Common. Damp flats and slopes, 5,000–7,000 ft. Mixed Conifer F., Red Fir F. Big Flat (*486a*). June–July.

Key to the families of the Monocotyledoneae

Perianth segments reduced, not petallike in color or texture.
- Fls. borne in the axils of chaffy scales and ± concealed by them.
  - Stems mostly triangular; lf. sheaths not split; fls. borne in the axil of a single bract.................. Cyperaceae p. 146
  - Stems mostly rounded; lf. sheaths usually split; fls. borne between 2 bracts .... Gramineae p. 153
- Fls. not concealed by chaffy scales.
  - Aquatic plants; fls. and lvs. floating or submersed. Sparganiaceae p. 176
  - Land plants of dry to wet places; fls. and lvs. not floating or submersed .... Juncaceae p. 162

At least the inner perianth segments well developed, petallike in color and texture.
- Ovary superior.
  - Fls. in a scapose umbel............ Amaryllidaceae p. 145
  - Fls. in racemes or panicles or solitary....... Liliaceae p. 167
- Ovary inferior.
  - Fls. regular .... Iridaceae p. 162
  - Fls. irregular .... Orchidaceae p. 172

AMARYLLIDACEAE

Perianth segments separate ....*Allium*
Perianth segments united below ....*Brodiaea*

Allium L.

Stems 30–70 cm tall......................................*A. validum*
Stems 3–20 cm tall.
- Lvs. falcate; ovary 3-crested ....*A. falcifolium*
- Lvs. not falcate; ovary 6-crested ....*A. campanulatum*

A. campanulatum Wats. Sierra Onion. Fig. 82. Scapose perennial from an ovoid bulb; stems 10–20 cm tall; lvs. 2, ca. equal to stem in length, 1–3 mm wide; umbel 10–40 fld., subtended by 2 ovate bracts 1–2 cm long; pedicels slender, 1–2 cm long; fls. rose to purplish, the segments 6–10 mm long, involute and somewhat keeled at the tips; ovary with 6, often purple crests at the summit; capsule ca. 3 mm high. Common. Open, dry slopes and flats,

5,000–7,000 ft. Mixed Conifer F., Red F. Canyon Cr. (*151, 783, 807a, 857*), Packers Pk. (*551*), Grizzly L. Basin (*896*), Caribou Basin (*776*). June–August.

A. falcifolium H. & A. Scythe-leaved Onion. Scapose perennial from a rounded bulb with a strong onion odor; stems 3–8 cm tall; lvs. 2, falcate, 7–12 cm long, 3–6 mm wide; umbel 10–25 fld., subtended by 2 ovate bracts 1–1.5 cm long; pedicels stoutish, 8–12 mm long; fls. pale rose to purple, the segments 6–9 mm long, involute at the tips; ovary with 3 central crests; capsule 4–5 mm long. Occasional. Dry, sandy flats and talus slopes, 6,000–7,000 ft. Red Fir F. Canyon Cr. (*161*), Packers Pk. (*547*). June–July.

A. validum Wats. Swamp Onion. Scapose perennial with rhizomes; stems 30–70 cm tall; lvs. 4–5, with an evident midrib below, 20–45 cm long; umbel many fld., subtended by 2–4 united bracts; fls. whitish to rose, the segments 7–9 mm long, acuminate; stamens and styles exserted; capsule 5–7 mm long, not crested. Occasional. Bogs and wet meadows, 6,500 ft. Red Fir F. Caribou Basin (*393*), Canyon Cr. (*827*). August.

Brodiaea Sm.

B. hyacinthina (Lindl.) Baker. Wild Hyacinth. Scapose perennial from deep-seated bulbs; stems slender, 3–4 dm tall; lvs. 1–3 dm long, linear; umbel 10–15 fld., subtended by 3–4 scarious bracts; pedicels slender, 1–2 cm long; fls. white with green midribs, the tube 4–5 mm long, the segments 6–8 mm long; stamens 6, alternating long and short, the filaments dilated and united at the base; capsule ca. 5 mm in diam. Occasional. Open or shaded meadows, 5,000 ft. Mixed Conifer F. Big Flat (*712*), along the South Fork of the Salmon R. (*1039*). August.

CYPERACEAE

Fls. perfect; achenes naked.. . . . . *Scirpus*
Fls. unisexual; achenes enclosed in a perigynium. . . . . *Carex*

Carex L.

Most of the species included here were determined by H. Leschke and J. T. Howell.

Spikelets gynaecandrous; infl. capitate to open.
  Perigynia prominently nerved dorsally and ventrally.
    Perigynia serrulate on the beak only, the margins of the body smooth . . . . *C. interior*
    Perigynia with winged, serrulate margins on the body and beak.
      Perigynia strongly nerved ventrally; female scales with narrow, hyaline margins above the middle. *C. abrupta*
      Perigynia with fewer and less prominent nerves

ventrally; female scales with wide, hyaline margins.
C. multicostata

Perigynia nerveless to very faintly nerved ventrally.

Perigynia 4–4.5 mm long.

Female scales brown with narrow, hyaline margins below the tip; perigynia plano-convex. . . . . *C. preslii*

Female scales hyaline with brown margins; perigynia flattened . . . . *C. proposita*

Perigynia 2.5–3.5 mm long.

Margins of perigynia smooth . . . . *C. integra*

Margins of perigynia serrulate.

Infl. conspicuously leafy-bracteate.
C. athrostachya

Infl. not conspicuously leafy-bracteate.
C. teneraeformis

Spikelets androgynous or unisexual; if some of the spikelets are gynaecandrous then the infl. is open and the spikelets are separate.

Styles mostly 3; achenes triangular.

Lowest bract of infl. leaflike and long sheathing.

Perigynia somewhat hispidulous toward the beak; lvs. 3–7 mm wide . . . . *C. ablata*

Perigynia glabrous; lvs. 1–2 mm wide. . *C. lemmonii*

Lowest bract of infl. sheathless or short sheathing.

Spikelets 1 per culm. . . . . *C. nigricans*

Spikelets 2–4 per culm.

Female scales dark purple-brown, the mid-nerve extended into a short awn; perigynia faintly nerved . . . . *C. spectabilis*

Female scales green with reddish brown margins, mostly awnless; perigynia strongly nerved.
C. rostrata

Styles 2; achenes mostly lenticular.

Perigynia narrowed to a slender, elongate, bidentulate beak.

Perigynia nerveless ventrally, the beak with serrulate margins . . . . *C. cusickii*

Perigynia few nerved ventrally, the margin of the beak entire. . . . . *C. jonesii*

Perigynia contracted to a short, entire beak or beakless.

Perigynia beakless . . . . *C. hassei*

Perigynia with a short beak.

Perigynia nerveless to faintly nerved.

Lowest bract of the infl. leaflike and equal to or longer than the infl. . . . . . *C. aquatilis*

Lowest bract of infl. bracteate and shorter than the infl. . . . . *C. scopulorum*

Perigynia strongly nerved . . . . *C. kelloggii*

C. ablata Bailey. American Cold-loving Sedge. Cespitose perennial from short, somewhat stout rootstocks; culms ca. 5 dm tall; lvs. 10–20 cm long, 3–7 mm wide; infl. ± crowded above, the spikelets widely separate below, spikelets 5–6, narrowly obovoid, 1–2 cm long, the terminal one male, the lower ones female; the lowest bract of the infl. leaflike and long sheathing; female scales shorter and ca. as wide as perigynia, purple-brown with a green midrib and broadly hyaline tips; perigynia 5–5.5 mm long, tapered to a bidentulate beak, nerved dorsally and ventrally, somewhat hispidulous toward the beak, the margins smooth; achenes light brown, ca. 1.5 mm long, mostly triangular with 3 styles but lenticular achenes with 2 styles can be found in the same spikelet. Found at one location. Stream bank, 6,000 ft. Red Fir F. Grizzly Cr. (*876*). August.

C. abrupta Mkze. Abruptly Beaked Sedge. Densely cespitose perennial from short rootstocks; culms ca. 5 dm tall; lvs. 10–20 cm long, 1–2 mm wide; infl. densely capitate, rounded to ovoid, 1–1.5 cm long; the spikelets gynaecandrous; female scales shorter and narrower than the perigynia, brown with a paler midnerve and narrow hyaline margins above the middle; perigynia ca. 4 mm long, contracted to a slender beak with a slit down one side, strongly nerved dorsally and ventrally, with narrowly winged, serrulate margins; achenes lenticular, light brown, ca. 1.5 mm long; styles 2. Found at one location. Stream bank, 6,000 ft. Mixed Conifer F. Grizzly Cr. (*875*). August.

C. aquatilis Wahl. Water Sedge. Cespitose perennial from elongate rootstocks; culms 4–5 dm tall; lvs. 10–30 cm long, 1.5–3 mm wide; infl. open; spikelets 3–6, linear, 1–3 cm long, the terminal 1 or 2 male, the lower female; female scales much shorter and narrower than the perigynia, dark purplish brown with a green midrib; perigynia ca. 2.5 mm long, abruptly contracted to a short, dark, entire beak, faintly nerved, the margins smooth; achenes lenticular, light brown, ca. 1 mm long; styles 2. Found at one location. Edge of lake, 6,000 ft. Mixed Conifer F. Josephine L. (*278*). August.

C. athrostachya Olney. Slender-beaked Sedge. Cespitose perennial from short rootstocks; culms 1.5–2 dm tall; lvs. 5–15 cm long, 1.5–2.5 mm wide; infl. densely capitate, ovoid, 8–10 mm long; the spikelets gynaecandrous; female scales shorter and narrower than the perigynia, brown with a green midrib; perigynia ca. 3.5 mm long, the beak slender and bidentate, faintly nerved dorsally, nerveless ventrally, with narrowly winged serrulate margins; achenes lenticular, pale, ca. 1.3 mm long; styles 2. Found at one location. Edge of ephemeral lake, 6,500 ft. Red Fir F. Dorleska Mine (*318*). August.

C. cusickii Mkze. Cusick's Sedge. Cespitose perennial from stout rootstocks; culms ca. 4 dm tall; lvs. 10–30 cm long, 1–4 mm wide; infl. paniculate, the spikelets androgynous, crowded on the branches; female scales light brown to hyaline, nearly concealing the perigynia; perigynia ca. 3 mm long, narrowed to a serrulate margined, bidentulate beak, nerved dorsally, nerveless ventrally, the margins of the body smooth, spongy at base; achenes lenticular, brown, ca. 1.5 mm long; styles 2. Found at one location. Bog, 5,000 ft. Mixed Conifer F. Big Flat (*234*). July–August.

C. hassei Bailey. Hasse's Sedge. Loosely cespitose perennial from long, slender rootstocks; culms 1–1.5 dm tall; lvs. 5–8 cm long, 1–3 mm wide; spikelets 4–6, narrow, 1–1.5 cm long, the terminal male, the lower ones female; the lowest bract of the infl. leaflike and long sheathing; female scales dark purple-brown with a broad green midrib, somewhat shorter and narrower than the perigynia to nearly concealing it; perigynia ca. 2.5 mm long, beakless, nerved, with smooth margins; achenes lenticular, light brown, ca. 2 mm long; styles 2. Occasional. Bogs, 5,000–6,000 ft. Mixed Conifer F. Mouth of Bullards Basin (*628a*). June–July.

C. integra Mkze. Smooth-beaked Sedge. Cespitose perennial from short rootstocks; culms 1–2.5 dm tall; lvs. 4–10 cm long, 1.5–2 mm wide; infl. loosely capitate, 1–1.5 cm long; spikelets gynaecandrous; female scales ca. as wide as the perigynia but a little shorter, brown with a green midrib, slightly keeled; perigynia 3 mm long, tapering to a short, bidentulate beak, nerved dorsally, nerveless to very faintly nerved ventrally, with very narrowly winged smooth margins; achenes lenticular, pale, ca. 1.3 mm long; styles 2. Found at one location. Damp, open meadow, 6,500 ft. Red Fir F. Yellow Rose Mine (*174*). June–July.

C. interior Bailey. Inland Sedge. Densely cespitose perennial from short rootstocks; culms 3–4 dm tall; lvs. 5–15 cm long, 1–2 mm wide, somewhat folded longitudinally; infl. ± capitate, ca. 1.5 cm long; spikelets gynaecandrous; female scales much shorter and somewhat narrower than the perigynia, brownish with broad hyaline margins; perigynia ca. 2.5 mm long, narrowed to a short, stout, bidentulate beak, nerved dorsally and ventrally, the margins of the body smooth, those of the beak minutely serrulate, achenes lenticular, dark brown, ca. 1.5 mm long; styles 2. Occasional. Bogs, 5,000–6,000 ft. Mixed Conifer F. Big Flat (*236*), Boulder Cr. Basin (*1426*). August.

C. jonesii Bailey. Jones' Sedge. Cespitose perennial from elongate rootstocks; culms ca. 2.5 dm tall; lvs. 5–10 cm long, 1–2 mm broad; infl. capitate, 8–12 mm long, the spikelets androgynous; female scales dark brown with a paler midrib, shorter than and as broad as the perigynia; perigynia ca. 3.5 mm long, tapered to a slender, bidentulate beak, somewhat spongy at the base, many nerved dorsally, few nerved ventrally, the margins entire; achenes lenticular, light brown, 1–1.3 mm long; styles 2. Found at one location. Bog, 5,000 ft. Mixed Conifer F. Big Flat (*531*). June–July.

C. kelloggii W. Boott. Kellogg's Sedge. Cespitose perennial from short rootstocks; culms 2–5 dm tall; lvs. 7–25 cm long, 1–2 mm wide, folded longitudinally toward the base; infl. open, spikelets linear, the terminal male, the lower ones female; female scales shorter or ca. as long as and narrower than the perigynia, dark purplish brown with a green midrib and narrow, hyaline margins toward the tip; perigynia 2–2.5 mm long, contracted to a short, entire beak, nerved dorsally and ventrally, the margins entire; achenes lenticular to somewhat angled, dark brown, ca. 1.3 mm long; styles 2. Found at one location. Edge of lake, 6,000 ft. Red Fir F. Boulder Cr. L. (*467, 1411*). August.

C. lemmonii W. Boott. Lemmon's Sedge. Cespitose perennial from elongate rootstocks; culms 3–4 dm tall; lvs. 3–10 cm long, 1–2 mm wide; spikelets 4–5, linear, 1–2 cm long, the upper crowded the lower separate, the terminal one male, the lower ones female; the lowest bract of the infl. leaflike and long sheathing; female scales shorter and somewhat narrower than the perigynia, brown with a green midrib; perigynia ca. 3 mm long, tapering to a narrow, bidentulate beak, few nerved, glabrous, the margins smooth; achenes triangular, light brown, ca. 1.5 mm long; styles 3. Found at one location. Bog, 5,800 ft. Mixed Conifer F. Mouth of Bullards Basin (*628*). July–August.

C. multicostata Mkze. Many-ribbed Sedge. Densely cespitose perennial from short rootstocks; culms 3–5 dm tall; lvs. 10–30 cm long, 1.5–3 mm broad; infl. densely capitate, ovoid, 1–2 cm long; spikelets gynaecandrous; female scales slightly shorter and narrower than the perigynia, light brown with wide hyaline margins; perigynia ca. 4 mm long, contracted to a narrow beak slit down the dorsal side, strongly nerved dorsally, the nerves fewer and less prominent ventrally, with narrowly winged, serrulate margins; achenes lenticular, light yellowish brown, ca. 2.3 mm long; styles 2. Occasional. Dry, open slopes, 5,500–6,500 ft. Mixed Conifer F., Red Fir F. Josephine L. Basin (*115*). June–July.

C. nigricans C. A. Mey. Blackish Sedge. Perennial from tough rootstocks; culms 1–2 dm high, ± arcuate; lvs. 5–10 cm long, mostly flat, 1–2.5 mm wide; spikelets solitary, 1–2 cm long, androgynous or entirely female; female scales dark brown and shiny, lanceolate, ca. 3 mm long; perigynia 2.5–3 mm long, greenish below, becoming dark toward the bidentulate tip, nerveless; styles 3. Found at one location. In mud on open granitic slope, 7,600 ft. Alpine Fell-field. Ridge west of upper Canyon Cr. (*1310*). August.

C. preslii Steud. Presl's Sedge. Densely cespitose perennial from short rootstocks; culms 2–3 dm tall; lvs. 10–20 cm long, 1–2 mm broad; infl. loosely capitate, 5–15 mm long; spikelets gynaecandrous; female scales shorter and narrower than the perigynia, brown with a pale midrib and narrow hyaline margins below the tip; perigynia 4–4.5 mm long, narrowed to a slender beak with a slit down the dorsal side, finely nerved dorsally, nerveless to very faintly nerved ventrally, with broadly winged serrulate margins; achenes lenticular, brown, 1.7–2 mm long; styles 2. Found at one location. Dry, rocky slope, 7,400 ft. Subalpine F. Head of Kidd Cr. (*1094*). August. Mr. Howell reports this as *C. preslii* nearing *C. pachystachya* var. *gracilis* (personal communication).

C. cf. *proposita* Mkze. Densely cespitose perennial from short rootstocks; culms 2–2.5 dm tall; lvs. 4–15 cm long, usually folded lengthwise at least toward the base, 1–2.5 mm broad; infl. ± open, 2–3 cm long; spikelets 4–6, gynaecandrous; female scales shorter and narrower than the perigynia, hyaline, with a green midrib and brownish margins; perigynia ca. 4 mm long, narrowed to a short beak with a slit down the ventral side, scarcely nerved dorsally, nerveless ventrally, with broadly winged, serrulate margins; achenes lenticular, pale, ca. 1.5 mm long; styles 2. Found at one location. Open, rocky slope, 7,500 ft. Subalpine F. Sawtooth Ridge west of Caribou Basin (*1077*). August. This is a tentative identification by Mr. Howell, the material collected being somewhat immature. *C. proposita* has been previously reported for Calif. only from the Sierra Nevada (Munz, 1959).

C. rostrata Stokes. Beaked Sedge. Loosely cespitose perennial from short, stoloniferous rootstocks; culms ca. 4 dm tall; lvs. 10–25 cm long, 2–6 mm wide; male spikelets 2–4, very narrow, ± crowded, ca. 4 cm long, female spikelets below, narrowly oblong, separate, 3–4 cm long; female scales shorter and much narrower than the perigynia, green with reddish brown margins; perigynia 4.5–6 mm long, narrowed to a long, slender, bidentate beak, nerved dorsally and ventrally, the margins smooth; achenes continuous with the persistent style base; styles 3. Occasional. Edge of lakes, 6,000 ft. Red Fir F. Josephine L. (*293*), Boulder Cr. L. (*1415*). August.

C. scopulorum Holm. Holm's Rocky Mountain Sedge. Loosely cespitose perennial from elongate rootstocks; culms 3–4 dm tall; lvs. 10–25 cm long, 2–4 mm wide; infl. somewhat crowded, spikelets 3–4, linear-oblong, the terminal one male, the lower ones female; female scales longer and narrower than the perigynia, dark purplish brown with a pale midrib; perigynia ca. 2.5 mm long, contracted to a short, entire beak, nerveless, the margins smooth; achenes lenticular, brown, ca. 1.7 mm long; styles 2. Found at one location. Bog, 6,500 ft. Red Fir F. Head of Union Cr. (*955*). August. This species has been previously reported for Calif. only from the Sierra Nevada, Mt. Lassen, and the Sweetwater and Warner Mts. (Munz, 1959).

C. spectabilis Dewey. Showy Sedge. Fig. 83. Loosely cespitose perennial from short rootstocks; culms 2–3 dm tall; lvs. 3–20 cm long, 2–4 mm wide; spikelets mostly 3–4, ± separate, the terminal male, 1.5–3 cm long, the female 1–2.5 cm long; female scales somewhat shorter or slightly exceeding and narrower than the perigynia, dark purplish brown with a pale midrib extended into a short awn; perigynia 3–4 mm long, extended into a short, bidentulate beak, faintly nerved, the margins smooth; achenes triangular, light brown, ca. 1.5 mm long; styles 3. Common. Wet banks and meadows,

6,000–8,500 ft. Red Fir F., Subalpine F., Alpine Fell-fields. Caribou Mt. (*344*), Boulder Cr. L. (*468*), below ice field on the north side of Thompson Pk. (*904*), Caribou Basin (*1047*), Sawtooth Ridge (*1066*). August–September.

C. teneraeformis Mkze. Sierra Slender Sedge. Cespitose perennial; culms ca. 4 dm tall; lvs. 10–15 cm long, 2–2.5 mm wide; infl. capitate, ovoid, ca. 1.5 cm long; spikelets gynaecandrous; female scales somewhat shorter and narrower than the perigynia, brown with a pale midrib and narrow hyaline margins; perigynia 2.5–3 mm long, narrowed to a slender bidentulate beak, nerved dorsally, nerveless ventrally, with broadly winged, serrulate margins; achenes lenticular, brown, ca. 1 mm long; styles 2. Found at one location. Bog, 6,500 ft. Red Fir F. Yellow Rose Mine (*951*). August.

Scirpus L.

Infl. capitate; bristles exserted beyond the scales . . . . *S. criniger*
Infl. paniculate, more open; bristles included within the scales.
  Style bifid; stamens 2; bristles stiff . . . . *S. microcarpus*
  Style trifid; stamens 3; bristles flexuous . . . . *S. congdonii*

S. congdonii Britton. Congdon's Bulrush. Perennial from stout rhizomes; stems slender, ± flattened, erect, 2–5 dm tall; lvs. well distributed, the blades 5–15 cm long, 3–7 mm wide; infl. a loose compound umbel, spikelets many; scales blackish with a pale midrib; bristles 4, flexuous, scarcely barbed, included within the scales; stamens 3; style mostly trifid; achenes pale brown, triangular, mucronate, ca. 1 mm long. Occasional. Stream banks and wet meadows, 6,000–7,000 ft. Red Fir F. Boulder Cr. Basin (*483, 1416*). August.

S. criniger Gray. Fringed Bulrush. Perennial from rhizomes; stems slender, triangular, erect, 2–6 dm tall; lvs. mostly basal, the blades 4–10 cm long, 2–4 mm wide; involucral lvs. 2–5, short, 4–10 mm long; infl. densely capitate, composed of 5–10 spikelets; scales brownish; bristles ca. 6, upwardly barbed, greatly exceeding the scales; stamens 3; styles trifid; achenes dark brown, triangular, oblong-obovate, mucronate, 2–3 mm long. Common. Bogs and wet meadows, 6,000–7,000 ft. Red Fir F. Head of Union Cr. (*996*). August.

S. microcarpus Presl. Small-fruited Bulrush. Perennial from stout rhizomes; stems rounded, erect, 4–5 dm tall; lvs. well distributed, the blades 10–15 cm long, 5–7 mm wide; infl. a loose, compound umbel, spikelets many; scales blackish with a green midrib; bristles 4, downwardly barbed, included within the scales; stamens 2; styles bifid; achenes whitish to pale brownish, lenticular, mucronate, ca. 1 mm long. Occasional. Bogs and stream banks, 6,500–7,000 ft. Red Fir F. Yellow Rose Mine (*948*). August.

GRAMINEAE

Spikelet with one fertile floret.
  Fertile floret subtended by an empty lemma which appears like a third glume; the fertile lemma indurate, shining, awnless. . *Panicum*
  Fertile floret not subtended by an empty lemma; the fertile lemma not as above.
    Florets 2, the lower perfect and awnless, the upper male, with a hooked awn . . . . *Holcus*
    Floret one.
      Lemma awned.
        Awn 3–4 cm long . . . . *Stipa*
        Awn 2 cm long or less.
          Awn arising from near base or middle of lemma. *Calamagrostis*
          Awn arising from the tip of lemma. *Muhlenbergia*
      Lemma awnless.
        Glumes stiffly ciliate on back, greatly compressed, awned . . . . *Phleum*
        Glumes not ciliate, not so compressed, awnless. *Agrostis*
Spikelet with 2 or more fertile florets.
  Infl. a spike, the spikelets alternating on opposite sides of the axis.
    Spikelets one at each node of the spike . . . . *Agropyron*
    Spikelets more than one per node.
      Spikelets 3 per node, the 2 lateral ones reduced and pedicelled . . . . *Hordeum*
      Spikelets 2 per node, both fertile and sessile.
        Spikes disarticulating at maturity; awns of glumes 2–7 cm long . . . . *Sitanion*
        Spikes continuous; awns of glumes less than 1 cm long or absent . . . . *Elymus*
  Infl. a raceme or panicle, the spikelets borne at the ends of the branches.
    Lemmas awned.
      Awn geniculate, dorsal.
        Lemmas keeled, awned from above the middle. *Trisetum*
        Lemmas convex, awned from middle or below. *Deschampsia*
      Awn straight, from the tip or bifid apex.
        Awn arising from the minutely bifid apex of lemma; spikelets 2–2.5 cm long . . . . *Bromus*
        Awn from tip of lemma, which is entire; spikelets ca. 1 cm long . . . . *Festuca*

Lemmas awnless.
Nerves of lemma parallel, prominent. . . . . *Glyceria*
Nerves of lemma converging toward summit, less prominent.
Spikelets 4–25 mm long; lf. blades 2–5 mm wide. *Melica*
Spikelets 3–9 mm long, mostly narrower; lf. blades 0.5–3 mm wide, often boatshaped at tips . . . . . . *Poa*

Agropyron Gaertn.

A. trachycaulum (Link.) Malte. Slender Wheat-grass. Culms tufted, erect, 5–10 dm tall; lf. blades 2–4 mm wide; spike slender, 10–25 cm long; spikelets only partially imbricate, 12–15 mm long, 3–5 fld.; glumes awnless, the first ca. 5 mm long, the second 6.5 mm long; lemmas 7–9 mm long, the awn 2 mm or less. Found at one location. Open meadow, 6,700 ft. Subalpine F. Ridge west of Bullards Basin (*969*). August.

Agrostis L.

Palea apparent, half as long as lemma. . . . . *A. alba*
Palea obsolete.
Panicle narrow and contracted, erect. . . . . *A. variabilis*
Panicle spreading, erect or drooping. . . . . *A. scabra*

A. alba L. Redtop. Perennial with elongate stolons; culms 5–12 dm tall; lf. blades 3–8 mm wide; panicle purplish to green, 1–3 dm long; glumes subequal, ca. 2.5 mm long, scabrous on keel; lemmas 2 mm long, lightly nerved; palea ca. half as long as lemma. Found at one location. Bog, 6,500 ft. Subalpine F. Yellow Rose Mine (*952*). August.

A. scabra Willd. Rough Hair-grass. Culms erect, tufted, 3–9 dm tall; lf. blades 1–3 mm wide; panicle 1.5–2.5 dm long, with capillary branches that spread or droop; glumes 2–2.5 mm long; lemmas 1.5–1.7 mm long; palea obsolete. Found at one location. Damp, shaded woods, 5,000 ft. Mixed Conifer F. Big Flat (*1039a*). August.

A. variabilis Rydb. Ross's Bent-grass. Culms slender, densely tufted, 1–2.5 dm tall; lf. blades to ca. 1 mm wide; panicle narrow with ascending branches, 2–6 cm long; glumes subequal, ca. 2 mm long; lemma 1.5 mm long; palea minute. Occasional. Rocky seeps, 6,000–7,000 ft. Red Fir F. Caribou Basin (*1074*). August.

Bromus L.

Panicle contracted; lemmas keeled on the back. . . . . *B. marginatus*
Panicle spreading; lemmas rounded on the back. . . . . *B. orcuttianus*

B. marginatus Nees. Large Mountain Brome-grass. Culms 6–12 dm tall; lf. blades 3–8 mm wide; panicle erect, narrow, ca. 1 dm long; spikelets ca. 2 cm long, 6–7 fld.; first glume ca. 8 mm long, the second ca. 10 mm long; lemmas ± pubescent, 12 mm long, with awns 4 mm long. Occasional. Dry, open slopes, 5,000–6,600 ft. Mixed Conifer F., Red Fir F. Yellow Rose Mine Tr. (*138*), Caribou Basin (*188*). June–July.

B. orcuttianus Vasey. Orcutt's Brome-grass. Culms erect, 7–15 dm tall; lf. blades 6–12 mm wide; panicle 1–1.5 dm long, the branches spreading; spikelets subterete, ca. 2.5 cm long, 5–7 fld.; first glume 5–8 mm long, the second 8–10 mm long; lemma ca. 10–12 mm long, the awn 5–7 mm long. Occasional. Dry, open slopes, 6,500 ft. Red Fir F. Yellow Rose Mine (*953*). August.

Calamagrostis Adans.

Awn shorter than the glumes, which are 3–5 mm long.
- Awn geniculate; callus bearing a tuft of hairs shorter than the lemma .... *C. koelerioides*
- Awn straight, delicate; callus bearing a copious tuft of hairs as long as lemma .... *C. canadensis*

Awn longer than the glumes, which are 6–8 mm long... *C. purpurascens*

C. canadensis (Michx.) Beauv. Blue-joint. Tufted perennial from creeping rhizomes; culms 6–7.5 dm high; lf. blades 3–5 mm wide; panicle lax, tending to be open, purplish; spikelets numerous, borne on slender pedicels; glumes nearly equal, narrowly lanceolate, 3–4 mm long, purplish; lemmas 2.5–3 mm long, the awn straight and delicate, shorter than glumes; callus bearing a copious tuft of hairs as long as lemma. Occasional. Wet banks and meadows, 5,500–6,500 ft. Mixed Conifer F., Red Fir F. Boulder Cr. Basin (*1417*), Canyon Cr. Lakes (*Alexander & Kellogg 5497*). July–August.

C. koelerioides Vasey. Tufted Pine-grass. Fig. 84. Culms ± tufted, 4–8 dm tall; lf. blades 2–3 mm wide; panicle narrow, 5–10 cm long, purplish; glumes subequal, 4–4.5 mm long; lemma ca. 4 mm long with a shorter geniculate awn from near the base; callus bearing a tuft of hairs 1–1.5 mm long. Common. Dry, open slopes, 5,000–6,500 ft. Mixed Conifer F., Red Fir F. Packers Pk. (*943*), Boulder Cr. Basin (*1410, 1420*). August.

C. purpurascens R. Br. in Richards. Purple Reed-grass. Culms tufted, erect, 4–10 dm tall; lf. blades 2–4 mm wide; panicle dense, narrow, 5–12 cm long, ± purplish; glumes slightly unequal, 6–8 mm long; lemma ca. as long, awned from near the base, the awn geniculate, ca. 8 mm long; callus with a tuft of hairs 2–2.5 mm long. Found at one location. Open, dry slope, 6,600 ft. Red Fir F. Yellow Rose Mine (*315*). August.

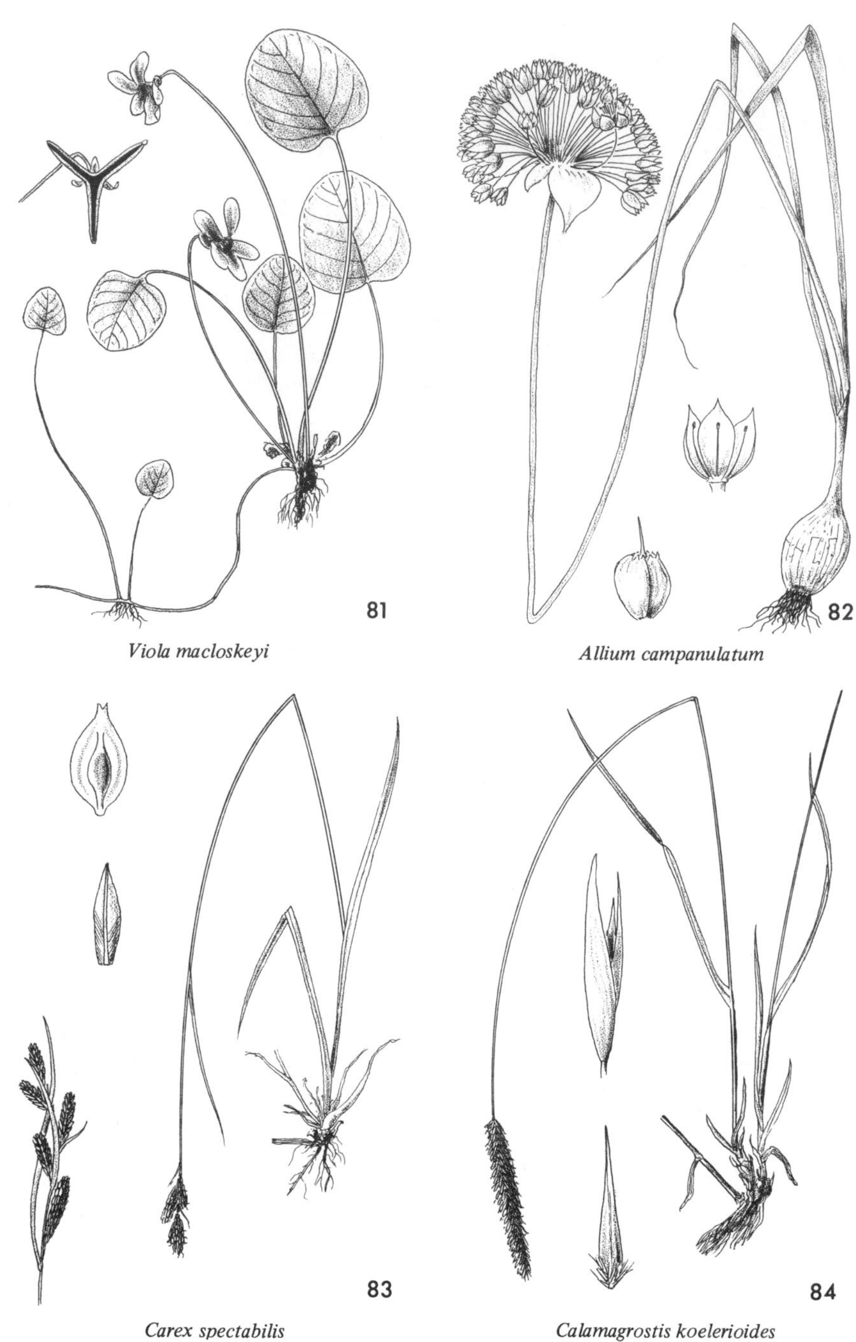

Viola macloskeyi

Allium campanulatum

Carex spectabilis

Calamagrostis koelerioides

Deschampsia Beauv.

D. atropurpurea (Wahl.) Scheele. Mountain Hair-grass. Tufted perennial, culms 3–4 dm high; lf. blades 3–8 cm long, 3–6 mm wide; infl. ± open, spikelets borne on capillary, scabrous pedicels; spikelets with 2 florets; glumes 4–5 mm long, purplish, 3-nerved; lemmas ca. 2 mm long, with a geniculate awn from the middle 2.5–3 mm long; callus with a tuft of hairs slightly shorter than the lemma. Occasional. Damp places, meadows and ridges, 7,000 ft. Red Fir F., Subalpine F. Caribou Basin (*1377*), ridge southwest of Boulder Cr. Basin (*1424*). August–September.

Elymus L.

E. glaucus Buckl. Western Rye-grass. Culms tufted, 5–6 dm high; lf. blades 3–8 mm wide; spike 7–10 cm long; spikelets mostly 2 per node, sessile; glumes 10 mm long, acuminate, strongly 2–4 nerved; lemmas 8–10 mm long, bearing a slender awn 15–22 mm long. Occasional. Meadows, 5,000–6,300 ft. Mixed Conifer F., Red Fir F. Caribou Mtn. (*1375*), Boulder Cr. Basin (*1431*). August–September.

Festuca L.

Awn ca. 2 mm long . . . . *F. idahoensis*
Awn 6–10 mm long . . . . *F. occidentalis*

F. idahoensis Elmer. Blue Bunch-grass. Culms slender, densely tufted, 3–10 dm tall; lvs. firm, 0.5–1 mm wide; panicle narrow, 1–2 dm long; spikelets ca. 8 mm long, 5–7 fld.; first glume ca. 3 mm long with one prominent nerve, the second 4 mm long, 5-nerved; lemmas 6–7 mm long, the awn ca. 2 mm long. Common. Dry, open slopes, 5,000–5,500 ft. Mixed Conifer F. Big Flat (*251*). August.

F. occidentalis Hook. Western Fescue. Culms slender, tufted, 4–10 dm tall; lvs. narrow, involute, ca. 1 mm wide; panicle narrow, 0.7–2 dm long; spikelets 3–5 fld., 6–10 mm long; glumes faintly nerved, 3 and 4 mm long; lemma 5–6 mm long, the awn slender, 6–10 mm long. Found at one location. Open slope, 5,500 ft. Mixed Conifer F. Below Josephine L. (*108*). June–July.

Glyceria R. Br.

G. elata (Nash) Hitchc. Tall Manna-grass. Clumped perennial, culms 7–10 dm high; lf. blades 4–8 mm wide, scabrous; ligules 3–4 mm long; spikelets many, in a loose panicle, purple tinged, 3–5 mm long; glumes purplish, the first 1 mm long, the second 1.5 mm long; lemmas 2–2.5 mm long, prominently 7-nerved. Occasional. Wet places, 5,000–6,500 ft. Mixed Conifer F., Red Fir F. Boulder Cr. Basin (*1290*). July–August.

Holcus L.

H. lanatus L. Velvet Grass. Velvety pubescent perennial; culms 3–10 dm tall; lf. blades 4–8 mm wide; panicle contracted, 8–15 cm long; spikelets ca. 4

mm long; glumes sharply keeled, the keels scabrous; lemmas smooth, shining; first floret perfect, awnless, the second male and with a hooked awn. Found at one location. Edge of bog, 5,000 ft. Mixed Conifer F. Big Flat (*373*). August.

Hordeum L.

H. brachyantherum Nevskii. Meadow Barley. Culms tufted, 2–7 dm tall; lf. blades 3–8 mm wide; spike 5–10 cm long; lateral spikelets pedicelled and sterile, the glumes slender, unequal, 10–12 mm long; central spikelet fertile, the lemma 7–10 mm long, the awn ca. 1 cm long; glumes ca. 9 mm long. Found at one location. Stream bank, 6,000 ft. Red Fir F. Grizzly Cr. (*878*). August.

Melica L.

Spikelets broadly V-shaped, reflexed on capillary pedicels. . . . *M. stricta*
Spikelets narrow, erect.
  First glume 3–4 mm long; rachilla swollen. . . . . *M. fugax*
  First glume ca. 5 mm long; rachilla normal. . . . . *M. bulbosa*

M. bulbosa Geyer ex Porter & Coulter. Western Melica. Culms 2–6 dm tall with or without a bulbous base; lf. blades 2–5 mm wide; panicle very narrow, 5–10 cm long; spikelets 6–25 mm long, 2–5 fld.; first glume ca. 5 mm long, the second 7–8 mm long, shorter than the spikelets; lemma strongly nerved, purplish below the tip, lemma of the first floret 6–11 mm long; the rudiment narrow, pointed, ca. 3 mm long. Occasional. Dry, open slopes, 6,500–7,500 ft. Red Fir F., Subalpine F. Yellow Rose Mine (*137*), Dorleska Summit (*301*). June–August.

M. fugax Bol. Small Onion-grass. Culms 2–6.5 dm tall from a bulbous base; lf. blades 2–4 mm wide; panicle narrow, with short, appressed to spreading branches, 7–18 cm long; spikelets 4–17 mm long, 4–5 fld., with a swollen rachilla; glumes with purple next to scarious margin and tip, the first 3–4 mm long; first lemma ca. 5 mm long, prominently nerved. Found at one location. Dry, open slope, 7,000 ft. Subalpine F. Dorleska Summit (*611*). July. In California this species has previously been reported only from the Sierra Nevada (Munz, 1959).

M. stricta Bol. Nodding Melica. Culms 2–5 dm high, densely tufted, purplish near the non-bulbous base; lf. blades 2–5 mm wide; panicle narrow, 0.3–2 dm long with appressed branches, the spikelets ± distant and reflexed; glumes scarious, 12–18 mm long; first lemma 8–16 mm long, faintly nerved. Found at one location. Steep talus slope, 7,000 ft. Subalpine F. The north side of Red Rock Mt. (*976*). August.

Muhlenbergia Schreb.

Plants 2–15 cm high; awn of lemma 1 mm long.......... *M. filiformis*
Plants 16–60 cm high; awn of lemma 2–15 mm long.
  Rhizomes prominent; second glume with a narrow tip, not 3-toothed .... *M. andina*
  Rhizomes not developed; second glume 3-toothed .... *M. montana*

M. andina (Nutt.) Hitchc. Hairy Muhlenbergia. Perennial from scaly rhizomes; culms slender, 3–4 dm tall; panicle narrow, ± interrupted, densely fld., 7–10 cm long; glumes narrow, ca. 2.5 mm long; lemma ca. 2.5 mm long, the awn 2–4 mm long; hairs at base of floret copious and ca. as long as lemma body. Found at one location. Dry, rocky slope, 5,500 ft. Mixed Conifer F. Below Josephine L. (*277*). August.

M. filiformis (Thurb.) Rydb. Slender Muhlenbergia. Delicate annuals or perennials; culms filiform, 0.2–1.5 dm high; lvs. narrow, 5–15 mm long; panicle narrow, few fld., pedicels appressed, 1–3 mm long; spikelets 2 mm long, glumes ovate, 0.7 mm long, lemmas minutely pubescent, bearing a short awn. Found at one location. Damp, sandy meadow, 5,700 ft. Red Fir F. Boulder Cr. Basin (*1428*). August.

M. montana (Nutt.) Hitchc. Mountain Muhlenbergia. Fig. 85. Culms slender, densely tufted, 1.6–6 dm high; lf. blades 1–2 mm wide; panicle narrow with appressed branches, 5–15 cm long; glumes scarious except for blackish coloration at the base, the first 1.5 mm long, the second slightly longer and 3-toothed; lemma 3–4 mm long, pilose below with short hairs, blackish at the tip with a slender awn 10–15 mm long. Occasional. On decomposed granite, 5,500–6,000 ft. Mixed Conifer F., Red Fir F. Josephine L. (*276*), Boulder Cr. Basin (*475*), Canyon Cr. (*769*). August.

Panicum L.

Spikelets glabrous; plants 3–12 cm tall.................. *P. capillare*
Spikelets pubescent; plants 10–20 cm tall............... *P. pacificum*

P. capillare L. Old-witch Grass. Annuals 3–12 cm tall with short flowering branches at the base; lf. blades and sheaths pubescent, the blades 2–4 mm wide and 1–2.5 cm long; panicle widely spreading, many fld., 3–5 cm long; spikelets ca. 2.5 mm long, sometimes purplish when young, glabrous; first glume ca. 1.3 mm long with a central prominent nerve, the second glume and sterile lemma alike, ca. 2.5 mm long, 5-nerved, acuminate; fertile lemma indurate, shining, 5-nerved, ca. 1.5 mm long. Found at one location. Dry lake bed, 5,000 ft. Mixed Conifer F. Big Flat (*364, 376*). August. Although the spikelets of these plants are comparable to those of typical *P. capillare*, the plants differ in the small size of the lvs., infl., and height of the stems.

P. pacificum Hitchc. & Chase. Pacific Panicum. Perennial, the culms erect,

1–2 dm tall; lf. blades and sheaths pilose, the blades 3–5 mm wide and 2–5 cm long; the panicle ± spreading, 2–3 cm long; spikelets pubescent, ca. 2 mm long, pale; first glume ca. 0.5 mm long, scarcely nerved, the second glume and sterile lemma similar, ca. 2 mm long, 7-nerved; fertile lemma indurate, shining, finely striate. Occasional. Dry slopes, 5,500–6,000 ft. Red Fir F. Boulder Cr. Basin (*460*). August.

Phleum L.

Awn of glumes 1.5–2 mm long; panicle ovoid.............. *P. alpinum*
Awn of glumes 0.7–1 mm long; panicle cylindrical.......... *P. pratense*

P. alpinum L. Mountain Timothy. Fig. 86. Culms slender, erect, 2–5 dm tall from a decumbent tufted base; lf. blades 3–6 mm wide; panicle dense, ovoid, ca. 1 cm wide and 1.5–3 cm long; glumes ca. 3 mm long, stiff ciliate, the awn 1.5–2 mm long; lemma ca. 2.5 mm long. Occasional. Damp open places, 6,000–7,000 ft. Red Fir F. Grizzly Cr. (*877*), Red Rock Mt. (*971*), Kidd Cr. (*1110*). August.

P. pratense L. Timothy. Culms tufted, slender, 4–10 dm tall, geniculate at base; lf. blades 5–8 mm wide; panicles dense, cylindric, ca. 5 mm wide and 3–12 cm long; glumes ca. 2.5 mm long, stiff ciliate, the awn 0.7–1 mm long; lemma ca. 1.5 mm long. Found at one location. Edge of bog, 5,000 ft. Mixed Conifer F. Big Flat (*372*). August.

Poa L.

Rhizomes present .... *P. compressa*
Rhizomes absent.
  Lemmas scabrous.
    Spikelets 5–8 mm long, the first lemma 4–6 mm long; culms 1.5–5 dm tall .... *P. epilis*
    Spikelets 7–9 mm long, the first lemma 5–6 mm long; culms ca. 2 dm tall .... *P. pringlei*
  Lemmas ± hairy on back, nerves or keel, at least toward the base.
    Many spikelets altered to dark purple bulblets with a slender appendage 2–3 cm long .... *P. bulbosa*
    Spikelets not altered to bulblets.
      Panicles ± spreading, the branches often at right angles to the axis .... *P. gracillima*
      Panicle contracted, the branches ascending.
        Culms ca. 3 dm tall .... *P. incurva*
        Culms 4–12 dm tall .... *P. canbyi*

P. bulbosa L. Bulbous Bluegrass. Culms 3–4 dm high from a ± bulbous base; lvs. 5–8 cm long, 1–2 mm wide; panicle dense, 5–8 cm long; unaltered spikelets 4–5 mm long, 5–6 fld.; glumes ca. 2.5 mm long; lemmas 2–3 mm long, ± silky-pubescent on keel and margins; bulblets ca. 3 mm long, dark purple, the bract extended into a slender, taillike appendage 2–3 cm long. Found at one location. Weedy area around ranch buildings, 5,000 ft. Mixed Conifer F. Big Flat (*1133*). June.

P. canbyi (Scribn.) Piper. Canby's Bluegrass. Fig. 87. Culms densely tufted, 4–12 dm tall; lvs. 0.5–2 mm wide; panicle narrow, ± compact, 6–15 cm long; spikelets 5–9 mm long, 3–6 fld.; lemmas 3.5–5 mm long, puberulent on lower half of back. Found at one location. Dry open slope, 6,500 ft. Red Fir F. Packers Pk. (*943a*). July–August.

P. compressa L. Canada Bluegrass. Culms from creeping rhizomes, 2–5 dm tall, flattened; lvs. 1–2 mm wide; panicle narrow, dense, 4–7 cm long; spikelets 4–6 fld., 3–4 mm long; lemmas ca. 2.5 mm long, purplish toward tip and with a moderately developed tuft of cobwebby hairs at the base. Found at one location. Dry open slope, 5,500 ft. Montane Chaparral. Near Yellow Rose Mine (*139*). June.

P. epilis Scribn. Mountain Bluegrass. Culms ± tufted, 1.5–5 dm tall; basal lvs. narrow, culm lvs. 1.5–3 mm wide; panicle ± condensed, ovoid to oblong, 2–6 cm long, purplish; spikelets 3–5 fld., 5–8 mm long; lower lemma 4–6 mm long, minutely scabrous. Occasional. Open slopes, 7,800 ft. Subalpine F., Alpine Fell-fields. Below the ice field on the north side of Thompson Pk. (*927*), summit of Packers Pk. (*1395*). July–August. Previously reported for Calif. only from the Sierra Nevada (Munz, 1959).

P. gracillima Vasey. Pacific Bluegrass. Culms densely tufted, 2–6 dm tall; lvs. mostly basal, ca. 1 mm wide; panicle 5–10 cm long, spreading, the branches often at right angles to the axis, the branchlets again spreading; spikelets 2–4 fld., 5–8 mm long; first lemma ca. 5 mm long, scarcely nerved, puberulent toward base. Found at one location. Open rocky slope, 7,400 ft. Subalpine F. Head of Kidd Cr. (*1095*). August.

P. incurva Scribn. & Will. Sandberg's Bluegrass. Culms slender, densely tufted, ca. 3 dm tall; lvs. mostly 1 mm wide; panicle narrow to somewhat spreading, with ascending branches; spikelets 2–4 fld., 5–7 mm long; lemmas 3–4 mm long, puberulent on lower half of nerves, purplish below the hyaline tip. Occasional. Open slopes, 5,500–7,000 ft. Red Fir F. Below Josephine L. (*114*), Grizzly L. (*881*). June–August.

P. pringlei Scribn. Pringle's Bluegrass. Culms slender, tufted, ca. 2 dm tall; lvs. 0.5–1 mm wide; panicle, narrow, ± condensed, purplish, 3–6 cm long; spikelets 7–9 mm long, 4–5 fld.; first lemma 5–6 mm long, keeled strongly 5-nerved, minutely scabrous on the nerves. Found at one location. Dry rocky ridge, 7,000 ft. Subalpine F. Dorleska Summit (*610*). June. Although difficult to separate from *P. epilis*, these plants seem to differ from it in having shorter culms, larger spikelets, and narrower lvs.

Sitanion Raf.

S. *hystrix* (Nutt.) J. G. Smith. var. *californicum* (J. G. Smith) F. D. Wilson. Bottle-brush Squirrel-tail. Culms loosely tufted, 1–5 dm tall; lvs. 1–5 mm wide; the spike 2–8 cm long, disarticulating at maturity; spikelets 2 per node, 2-few fld., one or both of them bearing a sterile floret at the base which appears like an extra glume; the glumes very narrow, the nerves extended into scabrous awns 2–7 cm long; the sterile floret similar to and ca. as long as the glumes; lemmas ca. 1 cm long with awns 2–10 cm long. Common. Dry meadows and open slopes, 5,000–7,300 ft. Mixed Conifer F., Red Fir F. Big Flat (*243*), Yellow Rose Mine (*316*), Willow Cr. Tr. (*427*), Packers Pk. (*942*). August. The interpretation here is that of Wilson (1963).

Stipa L.

S. *occidentalis* Thurb. Western Stipa. Culms tufted, 1.5–4 dm tall; lvs. 1–2 mm wide; panicle 1–2 dm long with ascending branches; glumes ca. 12 mm long, hyaline at the pointed tips; lemma ca. 7 mm long, pale brown, strigose; the awn 3–4 cm long, twice bent, plumose; callus well developed and bearded. Occasional. Dry flats and open slopes, 5,500–6,000 ft. Mixed Conifer F., Red Fir F. Yellow Rose Mine (*136*), Boulder Cr. Basin (*474*). June–August.

Trisetum Pers.

T. *spicatum* (L.) Richt. Downy Oat-grass. Culms slender, densely tufted, 1–4 dm tall; lvs. 1–3 mm wide; panicle narrow, dense, 3–6 cm long, pale; spikelets 4–6 mm long, 2-fld.; glumes 5–6 mm long, hyaline at margins and tips, the first 1-nerved, the second 3-nerved; lemmas ca. 5 mm long, the awn attached ca. one-third below the tip, 5–6 mm long and bent. Occasional. Shaded to open slopes, 7,500 ft. Red Fir F., Subalpine F. Caribou Mt. (*329*), Black Mt. (*453*). August.

IRIDACEAE

Sisyrinchium L.

S. *bellum* Wats. California Blue-eyed Grass. Tufted perennial from a short rootstock with fibrous roots; stems slender, 10–20 cm tall; lvs. mostly basal, linear, 4–8 cm long; infl. umbellate, from a spathelike bract; pedicels glabrous; perianth blue to purplish, 8–10 mm long, the segments aristulate; ovary glandular-pubescent; capsule 2–7 mm long. Found at one location. Open meadow, 5,000 ft. Mixed Conifer F. Along the South Fork of the Salmon R. (*101*). June–July.

JUNCACEAE

Lf. sheaths open . . . . *Juncus*
Lf. sheaths closed . . . . *Luzula*

Juncus L.

Stems 2–5 cm high; annuals . . . . *J. kelloggii*
Stems 10 or more cm high; perennials.

Auricles present at junction of sheath and blade.
  Anthers shorter than filaments . . . . *J. mertensianus*
  Anthers longer than filaments.
    Fls. borne singly, each one subtended by a pair of ovate bractlets.
      Lf. blade reduced to a bristle-like extension of the sheath 0.2–0.4 cm long; capsule retuse. *J. drummondii*
      Lf. blade not so reduced, 3–6 cm long; capsule mucronate . . . . *J. parryi*
    Fls. borne in compact heads. . *J. mertensianus* ssp. *gracilis*
Auricles lacking.
  Stamens 3 . . . . *J. ensifolius*
  Stamens 6 . . . . *J. orthophyllus*

J. drummondii E. Mey. Drummond's Rush. Stems tufted, slender, 15–30 cm high, from matted, fibrous roots; lf. blades reduced to a slender bristle 2–4 mm long, auricles ca. 0.5 mm long; fls. inserted singly, mostly 2–3 on a stem, each fl. subtended by a pair of ovate bractlets 2 mm long; perianth segments lanceolate-acuminate, 5–6 mm long, green with relatively broad purple-brown margins; stamens 6, filaments ca. 0.5 mm long, anthers ca. 1.2 mm long; capsule retuse, ca. as long as perianth, brown; seeds fusiform, ca. 0.5 mm long, brown, tailed at each end when young, the tail sometimes deciduous in age. Found at one location. Damp to drying places on open, granitic slope, 8,200 ft. Alpine Fell-field. North side of Thompson Pk. (*1305*). August. In California this species has been previously reported from the Sierra-Nevada and Siskiyou Co. (Munz, 1959).

J. ensifolius Wikstr. Three-stamened Rush. Perennial from slender, creeping rootstocks with many fibrous roots; stems compressed, 2–5 dm tall; lf. blades flattened, equitant, 7–12 cm long, 2–5 mm wide, auricles lacking; lowest bract of infl. sword-shaped, sometimes more than half as long as infl.; heads many, small, light to dark brown; perianth brown, 2.5–3 mm long; stamens 3, filaments ca. 1 mm long, anthers ca. 0.5 mm long; capsule 2.5–3 mm long, slightly longer than perianth, mucronate, dark brown, shiny; seeds broadly fusiform, ca. 0.5 mm long, brown, reticulate. Common. Bogs and damp places, 5,000–6,500 ft. Mixed Conifer F., Red Fir F. Big Flat (*235*), Canyon Cr. (*773*), Yellow Rose Mine (*950*). August.

J. kelloggii Engelm. Common Toad Rush. Annual with fibrous roots; stems 2–5 cm high, tufted; lvs. filiform, 1–2 cm long; fls. 2–3 per stem, in small, compact heads; perianth segments lanceolate, 2–2.5 mm long with hyaline margins and a green midrib; stamens 3, filaments 1–1.2 mm long, anthers ca. 0.5 mm long; capsule as long as perianth, mucronate, brown; seeds obovoid, brown, ca. 0.5 mm long. Found at one location. Edge of lake, 5,750 ft. Red Fir F. Boulder Cr. Basin (*1282*). August.

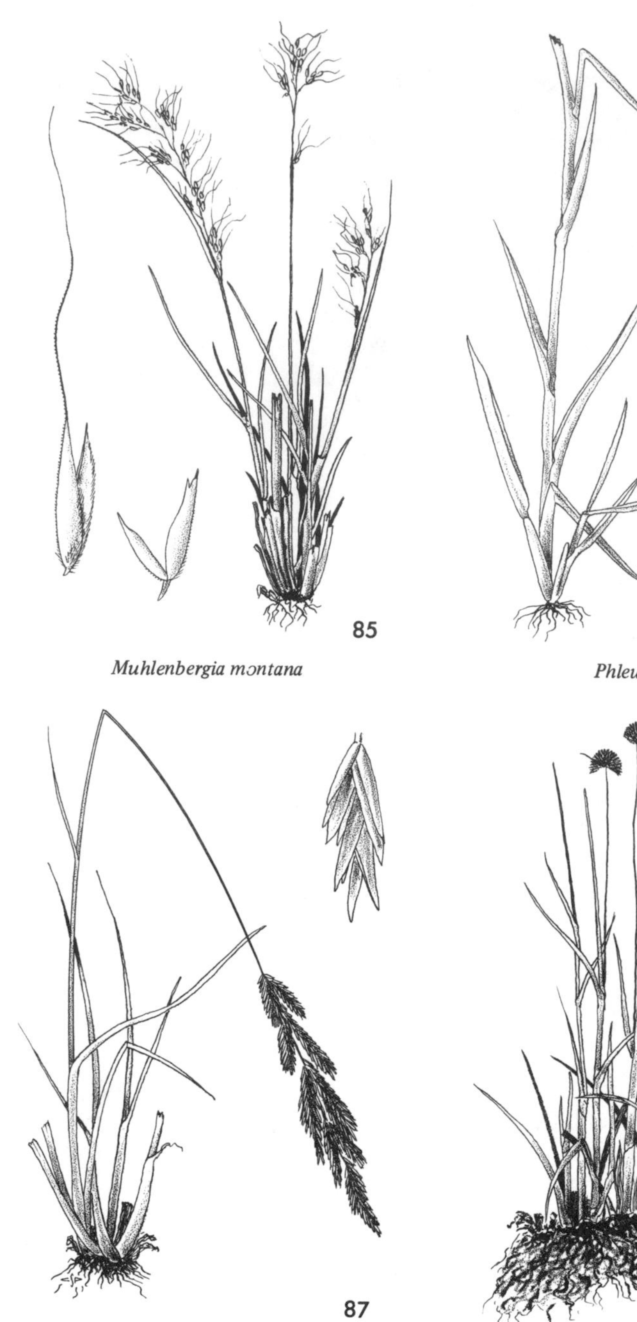

85 *Muhlenbergia montana*

86 *Phleum alpinum*

87 *Poa canbyi*

88 *Juncus mertensianus*

J. mertensianus Bong. Mertens' Rush. Fig. 88. Perennial from short rootstocks; stems tufted, slender, ± rounded 1.5–3 dm tall; lvs. somewhat flattened, 5–12 cm long, ca. 1 mm wide, auricles 1–2 mm long; heads solitary, many fld.; perianth dark purple-brown, ca. 3 mm long; stamens 6, filaments ca. 1 mm long, anthers ca. 0.5 mm long; capsule ca. as long as perianth, dark brown, retuse at apex, mucronate; seeds lance-ovoid, pale, ca. 0.5 mm long. Occasional. Wet, rocky, open slope, 7,600–8,500 ft. Alpine Fell-field. Below ice field on the north side of Thompson Pk. (*908*), Canyon Cr. (*1307*). August.

ssp. *gracilis* (Engelm.) F. J. Herm. Perennial from stout, creeping rootstocks; stems ± rounded, slender, 2–5 dm tall; lf. blades subrounded, septate, subfiliform, 3–15 cm long, auricles 1–2 mm long; lowest bract of infl. 1–1.5 cm long; heads many, in a loose, narrow panicle to 1 or 2 and crowded; perianth dark purple-brown to light brown, ca. 3 mm long; stamens 6, filaments ca. 0.5 mm long, anthers ca. 1.5 mm long; capsule brown, ca. as long as perianth, abruptly contracted into a short beak; seeds ovoid, broader toward one end, ca. 0.5 mm long, apiculate at both ends, dark brown, reticulate. Common. Bogs, 5,800–7,000 ft. Red Fir F., Subalpine F. Boulder Cr. L. (*466*), Yellow Rose Mine (*949*), Landers L. (*982*). August.

J. orthophyllus Cov. Straight-leaved Rush. Perennial from stout, creeping rootstocks; stems ± flattened, 3–4 dm tall; lvs. flattened, grasslike, short to almost as long as stem, 2–4 mm wide, auricles lacking; heads several, loosely paniculate; perianth 4–5 mm long, minutely roughened, brown with green midrib and scarious margins; stamens 6, filaments ca. 1 mm long, anthers ca. 1.5 mm long; capsule slightly shorter than perianth, mucronate; seeds obovoid, short-apiculate. Found at one location. Edge of lake, 5,800 ft. Red Fir F. Boulder Cr. L. (*465*). August.

J. parryi Engelm. Parry's Rush. Densely tufted perennial from numerous fibrous roots; stems rounded, slender, 1–2 dm tall; lf. blades grooved at base, rounded above, 3–6 cm long; fls. 1–4 per stem, inserted singly and subtended by 2 membranous, ovate bractlets; perianth light brown, 5–7 mm long; stamens 6, filaments ca. 0.5 mm long, anthers ca. 1.5 mm long; capsule light brown, ca. as long as perianth, mucronate; seeds light brown, reticulate, narrow-ovoid, ca. 0.7 mm long with a hyaline tail at each end 0.5–1 mm long. Common. Open rocky slopes, 5,500–7,800 ft. Red Fir F., Subalpine F., Alpine Fell-fields. Josephine L. Basin (*116*), below ice field on the north side of Thompson Pk. (928), Sawtooth Ridge (*1065*). June–August.

Luzula DC.

Fls. borne in compact heads . . . . *L. comosa*
Fls. borne singly at the ends of panicle branches.
    Panicle diffuse, the branches widely divergent . . . . *L. divaricata*

Panicle more compact, the branches drooping.
Perianth dark brown, the segments laciniate at the tips. *L. glabrata*
Perianth greenish to light brown, the segments ± entire at the tips . . . . *L. parviflora*

L. comosa E. Mey. Common Wood-rush. Tufted perennial with fibrous roots; stems 1.5–3 dm tall; lvs. 4–15 cm long, 2–6 mm wide, sparsely long pilose, especially at junction of sheath and blade; infl. umbellate, the rays 0.5–5 cm long, the fls. borne in several capitate clusters; perianth segments light brown, with broad hyaline margins, 1.5–3.5 mm long, entire at the tips; stamens 6, filaments and anthers ca. equal in length; capsule light brown, ca. as long as perianth; seeds red-brown, ca. 1 mm long with pale caruncle ca. 0.5 mm long. Common. Dry woods, 5,000 ft. Mixed Conifer F. Big Flat, along the South Fork of the Salmon R. (*407*, *675*, *1038*). July–August.

L. divaricata Wats. Forked Wood-rush. Tufted perennial with fibrous roots; stems 1–3 dm tall; lvs. 4–12 cm long, 2–5 mm wide, glabrous; infl. a diffuse panicle with widely divergent branches, the fls. borne singly at the ends of the branches; perianth brown, 2–2.5 mm long, the segments mostly entire at the tips; stamens 6, anthers ca. as long as filaments; capsule ca. as long as perianth, brown to greenish; seeds light brown, ca. 1 mm long, carunculate and with cobwebby hairs at the base. Common. Dry rocky slopes and ridges, 7,400–8,500 ft. Subalpine F., Alpine Fell-fields. Caribou Mt. (*328*, *348*), Black Mt. (*452*), north side of Thompson Pk. (*926*), Sawtooth Ridge (*1078*). August. This species has been previously reported for Calif. only from the Sierra Nevada (Munz, 1959).

L. glabrata (Hoppe) Desv. Smooth Wood-rush. Densely tufted perennial with short rootstocks; stems 1–2 dm tall; lvs. 3–6 cm long, 3–5 mm wide, sparsely long pilose at junction of blade and sheath; infl. a ± compact panicle, the fls. borne singly at the ends of the slender, drooping branches; perianth dark brown, 1.5–3 mm long, the segments laciniate-fringed at the tips; stamens 6, anthers ca. as long as filaments; capsule ca. as long as perianth, dark; seeds light yellowish, ca. 1.2 mm long, carunculate. Found at one location. Cracks in granite wall, 8,000 ft. Alpine Fell-field. Caribou Mt. (*349*). August.

L. parviflora (Ehrh.) Desv. Small-flowered Wood-rush. Fig. 89. Tufted perennial from short rootstocks; stems 3–5 dm tall; lvs. 4–15 cm long, 2–10 mm wide, sparsely long pilose at junction of sheath and blade; infl. a ± compact panicle, the fls. borne singly at the ends of the slender, drooping branches; perianth greenish to pale brown, 2–2.5 mm long, the segments ± entire at the tips; stamens 6, anthers ca. as long as filaments; capsule longer than perianth; seeds brown, ca. 1.5 mm long, with cobwebby hairs at base. Occasional. Damp, shaded, cool places, 6,000–7,000 ft. Red Fir F., Subalpine F. Kidd Cr. (*687*), Caribou Basin (*748*). July.

LILIACEAE

Lvs. basal.
  Perianth white to cream.
    Fls. 1–2 .... *Clintonia*
    Fls. several to many.
      Infl. a subcapitate panicle; stems viscid-pubescent. *Tofieldia*
      Infl. an elongate raceme; stems glabrous... *Schoenolirion*
  Perianth yellow to lavender or purple.
    Perianth yellow.
      Perianth segments 20–35 mm long; lvs. 2 ... *Erythronium*
      Perianth segments 6–8 mm long; lvs. several.. *Narthecium*
    Perianth not yellow.
      Fls. few; perianth lavender............... *Calochortus*
      Fls. many; perianth purple .... *Stenanthium*
Lvs. well distributed on the stem or immediately subtending the fls.
  Perianth segments 5–8 cm long........................ *Lilium*
  Perianth segments less than 3 cm long.
    Perianth dark purple-brown mottled with yellow... *Fritillaria*
    Perianth, at least the petals, white to green.
      Perianth segments 1–4 mm long.
        Plants from bulbs .... *Zigadenus*
        Plants from creeping rootstocks.......... *Smilacina*
      Perianth segments 8–20 mm long.
        Lvs. 3 in a whorl subtending the solitary fl... *Trillium*
        Lvs. many, alternate; fls. more than 1.
          Fls. many in a narrow to spreading panicle. *Veratrum*
          Fls. few, terminal or extra-axillary.
            Fls. extra-axillary, the peduncle twisted at junction with pedicel......... *Streptopus*
            Fls. terminal, peduncle absent, pedicel not twisted .... *Disporum*

Calochortus Pursh

C. nudus Wats. Naked Star Tulip. Fig. 90. Scapose perennial from a scaly bulb; stems 10–25 cm tall; lf. solitary, basal, 10–20 cm long, 2–11 mm wide; bracts of infl. 2, 1–2 cm long; fls. 1–3; sepals green to purplish, 1–2 cm long; petals pale lavender, 1.5–2 cm long, rounded and erose at apex, the gland bordered below by a broad, fringed membrane and with a few long hairs above; anthers and filaments lavender; capsule elliptic, winged, 15–20 mm long. Common. Meadows and damp places, 5,000–6,800 ft. Mixed Conifer F., Red Fir F. Big Flat (*100*, *587*), Josephine L. Basin (*272*), ridge west of Bullards Basin (*963*), Landers L. (*1012*). June–August.

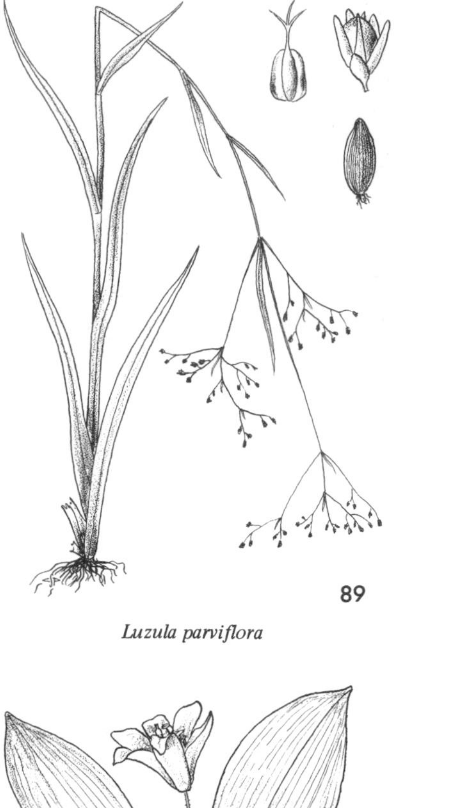

89

Luzula parviflora

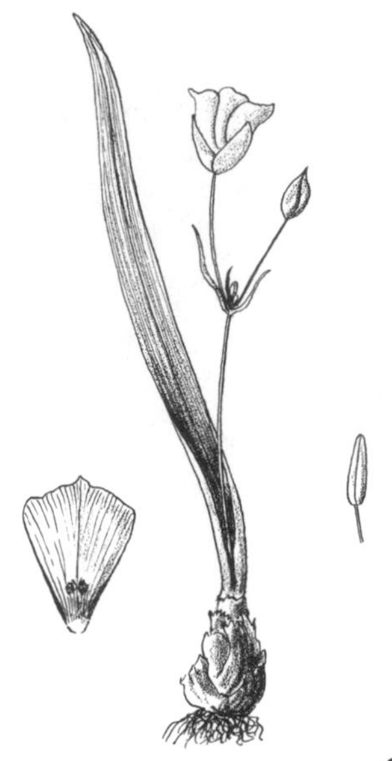

90

Calochortus nudus

91

Clintonia uniflora

92

Narthecium californicum

Clintonia Raf.

C. uniflora (Schult.) Kunth. Single-flowered Clintonia. Fig. 91. Scapose perennial from slender rootstocks; lvs. 2–3, obovate to oblanceolate, 6–15 cm long, 2–6 cm wide, sparingly pilose; peduncles 1-fld., 2–6 cm high, pilose; fls. white, the segments 1.5–2 cm long; fr. a berry, 8–12 mm long. Common. Shaded woods, 5,000 ft. Mixed Conifer F. Along the South Fork of the Salmon R. (*85, 702*), Canyon Cr. (*182*), old Caribou Basin Trail (*221*). June–July.

Disporum Salisb.

D. hookeri (Torr.) Nichols. var. *trachyandrum* (Torr.) Q. Jones. Hooker's Fairy Bell. Perennial from slender rootstocks; stems 1–4 dm long, branched and ± spreading above, pubescent throughout; 1–3 scarious bracts present on the lower stem; green lvs. borne on the upper branches, lanceolate to ovate with cordate-clasping bases, 1–6 cm long, 1.5–3 cm broad, ± hispid; fls. terminal, drooping, solitary or 2–3 in an umbel; pedicels 5–10 mm long; perianth segments greenish white, 9–12 mm long; anthers hispidulous; berry scarlet, ca. 8 mm long. Common. Shaded woods, 5,000 ft. Mixed Conifer F. Big Flat (*534*). June–July.

Erythronium L.

E. grandiflorum Pursh var. *pallidum* St. John. Yellow Fawn Lily. Perennial from a deep-seated, elongate corm; stems 2–4 dm tall, branched above; lvs. basal, mostly 2, oblanceolate, 10–20 cm long, 1.5–3 cm broad; fls. 2–5; perianth golden yellow, 2–3.5 cm long; styles plainly lobed; capsule obovoid, 2–3 cm long. Occasional. Damp, open meadows, 6,000–6,500 ft. Red Fir F. Head of Union Cr. (*647*). June.

Fritillaria L.

F. atropurpurea Nutt. Purple Fritillary. Perennial from a deep-seated bulb, rice grain bulblets sometimes present; stems 1–5 dm tall, simple, naked below; lvs. linear, 5–10 cm long, 2–6 mm wide; fls. 1–4, nodding on stout pedicels; perianth dark purple-brown mottled with yellow, 1–2 cm long; styles deeply cleft; capsule broadly obovoid, 10–17 mm long. Common. Dry slopes and open meadows, 6,500–7,500 ft. Red Fir F. Packers Pk. (*564*), Dorleska Mine (*615*). June–July.

Lilium L.

Perianth orange-yellow . . . . *L. kelleyanum*
Perianth white to pale lavender . . . . *L. washingtonianum*

L. kelleyanum Lemmon. Tiger Lily. Perennial from a deep-seated rhizomatous bulb; stems 4–20 dm tall, simple to branched above; lvs. lance-linear to oblanceolate, sessile, 4–10 cm long, 5–25 mm wide, alternate below, whorled above, margins planar; fls. 1–several, nodding; perianth segments reddish orange toward the tip, becoming orange-yellow with purple spots toward the base, reflexed ca. half way, ca. 5 cm long; anthers 4–6 mm long;

capsule oblong, 2.5 cm long. Common. Wet places, 5,500–7,000 ft. Mixed Conifer F., Red Fir F. Josephine L. (*280*), below Dorleska Mine (*625*), Caribou Basin (*721*), Canyon Cr. (*788*), Landers L. (*1009*). July–August.

L. washingtonianum Kell. Washington Lily. Perennial from a rhizomatous bulb; stems 8–15 dm tall; lvs. oblanceolate to obovate, sessile, ± undulate, 3–6 cm long, 6–20 mm wide, the lower alternate, the upper whorled; fls. 3–10, trumpet-shaped, fragrant; perianth segments white to pale lavender with purple spots, turning darker with age, 6–7 cm long; anthers 8–10 mm long; capsule oblong or obovoid, 2.5–3 cm long. Common. Dry, brushy or wooded slopes, 5,000–6,600 ft. Montane Chaparral, Red Fir F. Yellow Rose Mine (*296*), Caribou Basin (*720*). July–August.

Narthecium Huds.

N. californicum Baker. California Bog-asphodel. Fig. 92. Perennial from creeping rootstocks; stems slender, 1–3 dm tall; lvs. basal, grasslike, equitant, 2–15 cm long, 1–4 mm wide, with prominent parallel veins; racemes several fld.; pedicels 6–8 mm long; perianth segments greenish without, bright yellow within, 4–7 mm long; filaments densely yellow-woolly; capsule ovoid-lanceolate, 8–12 mm long. Common. Bogs and seeps, 5,000–7,000 ft. Mixed Conifer F., Red Fir F. Caribou Mt. (*402*), Big Flat (*530*), Landers L. (*1013*). June–August.

Schoenolirion Durand

S. album Durand. White-flowered Shoenolirion. Fig. 93. Perennial from a stout bulb; stems 3–6 dm tall, glabrous; lvs. basal, 10–30 cm long, 5–12 mm wide with a ± evident pale midrib; racemes several to many fld.; pedicels ca. 2 mm long; perianth segments white with a dark central vein, ca. 5 mm long, persistent and becoming scarious; capsule ovoid, 6–7 mm long. Common. Bogs and seeps, 5,000–7,000 ft. Mixed Conifer F., Red Fir F. Canyon Cr. (*153, 774*), Caribou Basin (*203*), Big Flat (*230*), head of Union Cr. (*1005*). July–August.

Smilacina Desf.

Fls. in panicles, numerous; lvs. ovate to oblong, broadest near middle.
S. racemosa var. *amplexicaulis*

Fls. in racemes, few–several; lvs. lanceolate, broadest toward base.
S. stellata

S. racemosa (L.) Desf. var. *amplexicaulis* (Nutt.) Wats. Western Solomon's Seal. Perennial from creeping rootstocks; stems 2–3 dm tall; lvs. ovate to oblong with sessile, clasping bases, 5–9 cm long, 1.5–4 cm wide, well distributed on the stem; panicle many fld.; pedicels 0.5–1 mm long; perianth white, the segments 1–1.5 mm long; stamens 2–2.5 mm long; berry ca. 5 mm long, mostly red. Found at one location. Shaded slope, 6,000 ft. Red Fir F. Yellow Rose Mine Trail (*569*). June–July.

S. stellata (L.) Desf. Nuttall's Solomon's Seal. Fig. 94. Perennial from creeping rootstocks; stems 3–5 dm tall; lvs. lanceolate with sessile, ± clasping bases, 4–12 cm long, 0.5–3.5 cm wide; racemes few–several fld.; pedicels 6–15 mm long; perianth white, the segments 2.5–3.5 mm long; stamens ca. 2 mm long; berry 5–10 mm long, red-purple, becoming black. Common. Shaded woods, 6,000–6,500 ft. Red Fir F. Canyon Cr. (*154*, *835*), Union L. (*643*). June–August.

Stenanthium Gray

S. occidentale Gray. Western Stenanthium. Perennial from a bulb 1–2.5 cm long; stems solitary, sometimes branched above, 3–6 dm high; lvs. mostly, 10–35 cm long, 3–22 mm wide; infl. 5–30 cm long; fls. nodding on slender pedicels 10–30 mm long, bracts lanceolate, 10–25 mm long; perianth campanulate, 10–20 mm long, the segments becoming reflexed at tips, purple-green to purple-yellow; capsule narrowly ovoid, 15–20 mm long; styles separate, 2–3 mm long. Occasional. Damp banks, 5,000–6,300 ft. Mixed Conifer F., Red Fir F. Stuart Fork (*J. O. Sawyer 2359*), Union Cr. (*H. M. Hall 8626*). July–August.

Streptopus Michx.

S. amplexifolius (L.) DC. var. *denticulatus* Fassett. Clasping-leaved Twisted-stalk. Perennial from a stout rootstock with thick, fibrous roots; stems 4–7 dm tall, sparingly stiff-pubescent below; lvs. ovate-lanceolate, green above, glaucous beneath, 6–12 cm long, 3–6 cm wide, with clasping bases and minutely denticulate margins; fls. several on slender, twisted, extra-axillary peduncles; perianth greenish to white with purple on the back, the segments somewhat spreading, 1–1.5 cm long; berries reddish orange, 1–1.6 cm long. Occasional. Shaded stream banks, 6,000–6,500 ft. Red Fir F. Kidd Cr. (*690*, *1102*). July–August.

Tofieldia Huds.

T. glutinosa (Michx.) Pers. ssp. *occidentalis* (Wats.) C. L. Hitchc. Western Tofieldia. Scapose perennial from slender rootstocks bearing many fibrous roots; stems viscid-pubescent, 2–5 dm tall; lvs. basal, equitant, grasslike, 4–30 cm long, 1–6 mm wide; infl. a subcapitate, many fld. panicle; perianth greenish white to somewhat purplish, 3–4 mm long; capsule 5–9 mm long, turning red with age. Common. Bogs and seeps, 5,000–7,000 ft. Mixed Conifer F., Red Fir F., Subalpine F. Big Flat (*369*), Canyon Cr. (*806*), Grizzly L. (*879*), head of Union Cr. (*997*). July–August.

Trillium L.

T. ovatum Pursh ssp. *oettingeri* Munz & Thorne. Western Wake-robin. Perennial from short rootstocks; stems simple, erect, 12–30 cm tall; lvs. in a whorl of 3 subtending the solitary fl., broadly ovate, obtuse at the apex to strongly cuspidate, 5–10 cm long, 2.5–8 cm wide, sessile to having short petioles 2–3 mm long; pedicels 1–2 cm long; sepals green, lanceolate, 12–15 mm long; petals white, turning purple with age, lanceolate, 10–20 mm long, 3–5 mm wide; anthers 2–5 mm long; berry 1–1.5 cm long, only scarcely winged. Common. Damp, shaded woods, 5,000–6,000 ft. Mixed Conifer F., Red Fir F. Canyon Cr. (*60, 856*), along South Fork of the Salmon R. (*102*), Josephine L. (*597*), Kidd Cr. (*691*). June–July. These plants differ from *T. ovatum* in having shorter sepals, petals, and anthers, and in most of the lvs. being petiolate. The nomenclature here is that proposed by Munz & Thorne (1973).

Veratrum L.

Fls. white; lower panicle branches ascending to erect. . . *V. californicum*
Fls. green; lower panicle branches drooping. . . . . *V. viride*

V. californicum Durand. California False Hellebore. Stout perennial from thick rootstocks; stems 1–2 m tall, tomentose above; lvs. ovate to broadly elliptic, 10–30 cm long, 10–15 cm wide with clasping bases; infl. a large, spreading, many fld. panicle, the lower branches ascending to erect; fls. white, the segments 10–15 mm long with a blue-green V-shaped gland at the base within; capsule 2–3.5 cm long. Common. Wet meadows, 5,000–7,000 ft. Mixed Conifer F., Red Fir F. Dorleska Mine (*307*), Ward L. (*448*), Big Flat (*708*), Grizzly Cr. (*867*). July–August.

V. viride Ait. Green False Hellebore. Stout perennial from a thickened rootstock; stems 1–2 m tall, tomentose above; lvs. ovate to lanceolate, 15–25 cm long, 6–10 cm wide with clasping bases; infl. a narrow, many fld. panicle, the lower branches drooping; fls. green, the segments 8–10 mm long with a V-shaped gland at the base within; capsule 2–2.5 cm long. Occasional. Damp meadows and slopes, 7,000 ft. Red Fir F., Subalpine F. Canyon Cr. (*824*), Caribou Basin (*1043*). August.

Zigadenus Michx.

Z. venenosus Wats. Death Camass. Perennial from oblong-ovoid bulbs; stems slender, 2–4 dm tall; lvs. linear, grasslike, 10–25 cm long, 2–5 mm wide; infl. a many fld. raceme; pedicels slender, 1.5–2.5 cm long; fls. white, the segments 3–4 mm long; stamens slightly longer; capsule cylindric, 1–1.5 cm long. Found at one location. Wet meadow, 5,100 ft. Mixed Conifer F. Boulder Cr. Basin (*179*). June–July.

ORCHIDACEAE

Lvs. reduced and scalelike; plants not green. . . . . *Corallorhiza*
Lvs. well developed; plants green.
    Fls. 1–3, large and showy.
        Lf. 1, basal; fl. 1. . . . . *Calypso*
        Lvs. 4–7, cauline; fls. 1–3. *Cypripedium*
    Fls. several–many in elongate spikes or racemes.

Lip projected backward into a narrow, elongate spur. *Habenaria*

Lip lacking such a spur.
Lvs. 2, subopposite .... *Listera*
Lvs. 3–several.
Spike twisted; lip not inflated; lvs. linear. . *Spiranthes*
Spike not twisted; lip inflated into a scrotiform sac; lvs. ovate........................ *Goodyera*

Calypso Salisb.

C. bulbosa (L.) Oakes. Calypso. Scapose perennial from a rootstock with coralloid roots; scapes 10–15 cm high, with 2 sheathing bracts; lf. basal, ovate, 5–6 cm long, petioles 1–2 cm long; sepals and petals lanceolate, ca. 2 cm long, purple, 3-nerved; lip saccate, the pouch 1.5 cm long, white, streaked and spotted with red-purple; capsule ca. 1 cm long. Found at one location. Forest floor, 5,500 ft. Mixed Conifer F. Packers Pk. (*1147*). June.

Corallorhiza Chat.

Lip with prominent longitudinal stripes.................. *C. striata*
Lip without prominent longitudinal stripes.
Spur wholly attached to ovary; column 4–5 mm long. . *C. maculata*
Spur free at tip for ca. 1 mm; column 6–8 mm long. . *C. mertensiana*

C. maculata Raf. Spotted Coral-root. Scapose perennial from short, branched rhizomes; stems 2–4 dm tall, reddish brown to purple; lvs. mostly 3, reduced, long-sheathing, brownish; fls. many; sepals 5–10 mm long, purplish with 3 darker nerves; spur wholly attached to ovary; petals similar to sepals; lip white, spotted and veined with purple, 6–8 mm long, with 2 small, lateral lobes and 2 central ridges toward the base; column yellow, 4–5 mm long; capsule nodding, 1–2 cm long. Common. Dry, shaded woods, 5,000–7,000 ft. Mixed Conifer F., Red Fir F. Along the South Fork of the Salmon R. (*83, 84, 111*), Caribou Basin (*741*), Caribou Mt. (*324*), Canyon Cr. (*836*). June–August.

C. mertensiana Bong. Merten's Coral-root. Scapose perennial from rhizomes; stems 2–3.5 dm tall, brown to purplish; lvs. 3, brownish, reduced, long-sheathing; fls. several to many; sepals 5–10 mm long, purplish, faintly nerved; spur free at tip for ca. 1 mm; petals similar to sepals; lip pale to purplish, faintly nerved to somewhat spotted with purple, ca. 7 mm long, entire to somewhat lobed laterally, clawed at base; column slender, 6–8 mm long; capsule mostly ascending, 1.2–1.6 cm long. Occasional. Dry, shaded woods, 5,000 ft. Mixed Conifer F. Canyon Cr. (*181*), Kidd Cr. Trail (*676*). June–July.

C. striata Lindl. Striped Coral-root. Scapose perennial from short rhizomes; stems pale, 1.5–4 dm tall; lvs. 3, reduced, whitish to purplish, long-sheathing;

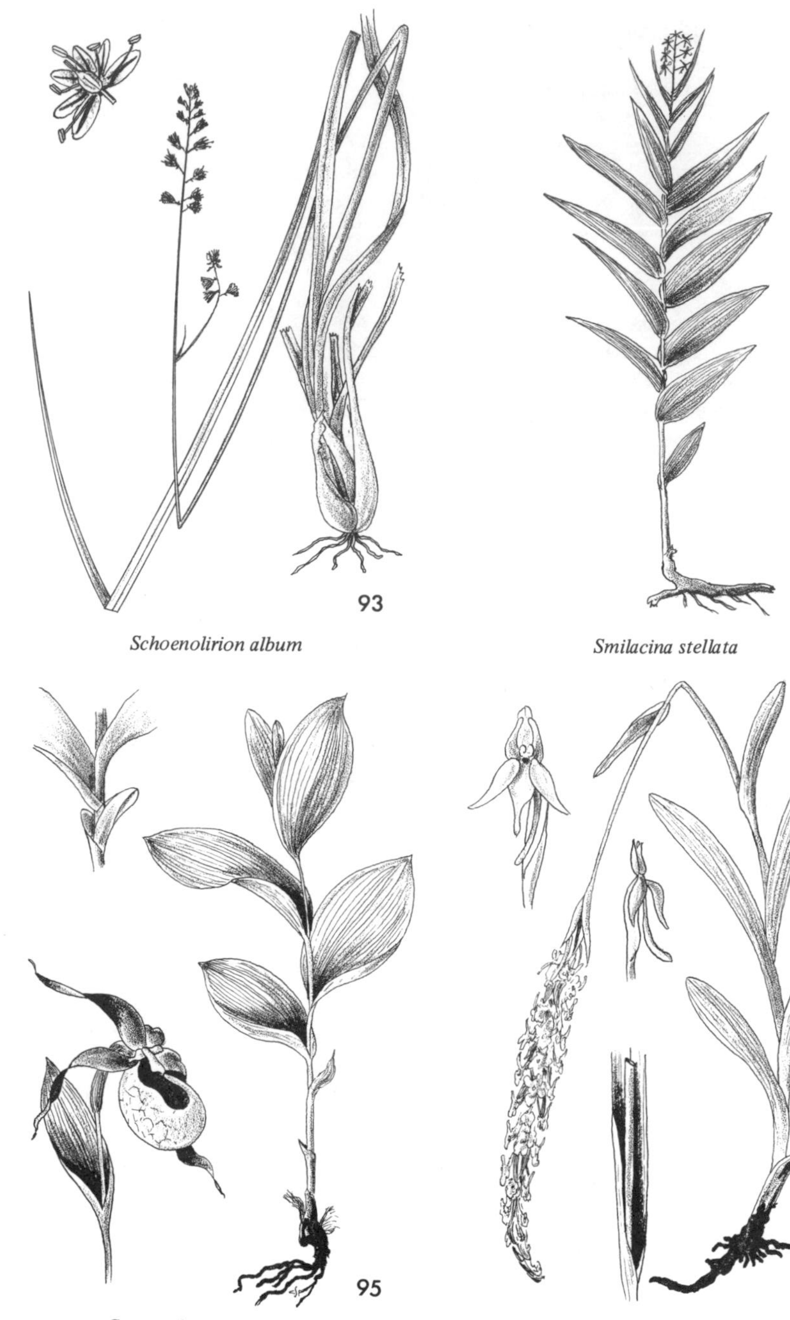

Schoenolirion album

Smilacina stellata

Cypripedium montanum

Habenaria dilatata var. leucostachys

fls. many; sepals 10–15 mm long, pale yellow, strongly 3–5 nerved; petals similar to sepals; lip white with prominent purple veins, ca. 10 mm long; column 3–5 mm long; capsule 1.2–2 cm long. Occasional. Dry, shaded woods, 5,000 ft. Mixed Conifer F. Big Flat (*580*). June–July.

Cypripedium L.

C. montanum Dougl. ex Lindl. Mountain Lady's Slipper. Fig. 95. Perennial from a short rootstock with tough, fibrous roots; stems simple, 2–4 dm tall, glandular-pubescent; lvs. 4–7, broadly ovate to elliptic, 6–13 cm long, ± glandular-puberulent; fls. 1–3, with a distinctive fragrance; sepals dark brownish purple, narrow, ca. 4 cm long; petals somewhat smaller than sepals and greenish to darker; lip white with purple veins, 2–3 cm long, inflated and saccate; capsule ascending, 2 cm long. Rare. Among shrubs on forest floor, 5,000 ft. Mixed Conifer F. Big Flat (*602*). June–July.

Goodyera R. Br.

G. oblongifolia Raf. Rattlesnake Plantain. Perennial from fleshy rootstocks with fibrous roots; stems 2–4 dm tall, glandular-pubescent; lvs. basal, the blades ovate, 4–5 cm long, white-veined, petioles short, winged; fls. many; sepals green, glandular-pubescent, ca. 6 mm long, the lateral pair free, the dorsal one fused with the petals and forming a galea; lip whitish, saccate; capsule ca. 1 cm long. Common. Dry woods, 5,000–5,500 ft. Mixed Conifer F. Grizzly Cr. (*861*). August.

Habenaria Willd.

Lvs. mostly well distributed along the stem; sepals 3-nerved.
  Fls. white . . . . *H. dilatata* var. *leucostachys*
  Fls. green . . . . *H. sparsiflora*
Lvs. basal; sepals 1-nerved . . . . *H. unalascensis*

H. dilatata (Pursh) Hook. var. *leucostachys* (Lindl.) Ames. Boreal Bog Orchid. Fig. 96. Perennial from fleshy roots; stems 2–5 dm tall; lvs. linear-lanceolate, long-sheathing, 4–15 cm long, well distributed up the stem; fls. many, crowded in a dense spike; perianth white; sepals 3-nerved, 4–6 mm long; petals similar to sepals; lip 7–8 mm long, dilated at base; spur filiform, curved, 10–15 mm long; capsule 1–1.5 cm long. Common. Bogs, stream banks, and other wet places, 5,000–7,000 ft. Mixed Conifer F., Red Fir F. Big Flat (*141*, *528*), Canyon Cr. (*847*), Red Rock Mt. (*970*). June–August.

H. sparsiflora Wats. Sparsely Flowered Bog Orchid. Perennial from fleshy roots; stems 3–5 dm tall; lvs. lanceolate to lance-oblong, 4–9 cm long, sessile with clasping bases, well distributed on the stem; fls. several to many; perianth green; sepals 3-nerved, the lateral 7–8 mm long, dorsal sepal shorter and broader; petals ca. 6 mm long; lip linear, 7–10 mm long; spur paler and ca. as long as lip; capsule 1.2–1.5 cm long. Occasional. Bogs and wet, shaded places, 6,500 ft. Red Fir F. Kidd Cr. (*685*), Caribou Basin (*723*). July. Material from

Canyon Cr. (*848*) has narrower lvs. mostly toward the base of the stem and somewhat smaller fls., but the sepals are 3-nerved and the spur is at least as long as the lip.

H. unalascensis (Spreng.) Wats. Alaska Piperia. Perennial from fleshy roots; stems 2–6 dm tall; lvs. linear-oblanceolate, 5–12 cm long, basal; fls. many in a long, narrow spike; perianth green; sepals 1-nerved, ca. 1.5 mm long; petals similar to sepals; lip broader, ca. 2 mm long; spur filiform, curved, slightly longer than the lip; capsule 1–1.5 cm long. Common. Dampish to dry, rocky slopes, 6,000–7,000 ft. Red Fir F. Caribou Basin (*742*), Canyon Cr. (*803*), Grizzly L. Basin (*886*). July–August.

Listera R. Br.

L. convallarioides (Sw.) Torr. Broad-lipped Twayblade. Delicate perennial with fibrous roots; stems 1–2 dm tall, pubescent above the lvs.; lvs. 2, subopposite, broadly ovate to nearly round, with sessile, clasping bases, 3–6 cm long, 2.5–4.5 cm wide; fls. several; perianth green; sepals and petals yellow-green, 4–5 mm long, reflexed in flowering; lip greenish, clawed, cuneate, 10–12 mm long, notched at apex; capsule glandular-puberulent, ca. 6 mm long. Common. Damp, shaded places, 5,500–6,500 ft. Red Fir F. Caribou Basin (*724*), Canyon Cr. (*830*), Grizzly Cr. (*864*). July–August.

Spiranthes Rich.

S. romanzoffiana Cham. & Schlecht. Hooded Ladies' Tresses. Perennial from clustered, tuberous roots; stems 1–2 dm tall, glandular-puberulent in infl.; lvs. linear to linear-lanceolate, 3–8 cm long; fls. several–many in a crowded, twisted spike; perianth greenish white; sepals and petals 6–8 mm long; lip ca. 6 mm long, abruptly narrowed below the dilated apex; capsule ellipsoid to ovoid. Common. Damp sandy flats, 5,500–6,500 ft. Red Fir F. Caribou Basin (*391*), Boulder Cr. L. (*469*), Canyon Cr. (*768*). August.

SPARGANIACEAE

Sparganium L.

S. angustifolium Michx. Narrow-leaved Bur-reed. Aquatic perennial from creeping rhizomes; stems slender, floating, 3–5 dm long; lvs. floating, nearly as long as the stems, 2–4 mm wide; fls. borne in several dense, rounded heads, the male heads above the female; fls. unisexual, the perianth segments reduced to a few, chaffy, elongate scales; achenes fusiform, 5–6 mm long, stipitate, the beak ca. 1 mm long. Occasional. In shallow water at edges of lakes, 5,600 ft. Red Fir F. Canyon Cr. (*485*), Boulder Cr. Basin (*1281*). August.

List of Additional Species

collected or observed above 5,000 ft.

Much of the work in compiling this list was done at the California Academy of Sciences Herbarium, San Francisco. I am especially grateful to John Thomas Howell for his assistance on numerous occasions. The list is arranged alphabetically by family and includes genus, species, habitat if known, and collector's number which may be identified as follows: those preceded by the letter *H* are Mr. Howell's; those without a letter are the author's; other collectors are named.

AMARYLLIDACEAE

Allium parvum Kell. Loose, rocky soil, 6,500 ft. Packers Pk. (*1153*).

ARISTOLOCHIACEAE

Asarum hartwegii Wats. Caribou Gulch (*H13,542*).

BETULACEAE

Alnus tenuifolia Nutt. 5,000 ft. South Fork of Salmon R. near Big Flat (*H13,350*).

BORAGINACEAE

Cryptantha simulans Greene. 5,000 ft. South Fork of Salmon R. near Big Flat (*H13,183*).

C. torreyana (Gray) Greene. 5,000 ft. South Fork of Salmon R. near Big Flat (*H13,182*).

CAPRIFOLIACEAE

Sambucus caerulea Raf. 5,000 ft. South Fork of Salmon R. near Big Flat (*H13,358*).

Symphoricarpos mollis Nutt. in T. & G. 5,000 ft. South Fork of Salmon R. near Big Flat (*H13,362*).

S. rivularis Suksd. 5,000 ft. South Fork of Salmon R. near Big Flat (*H13,188*).

CARYOPHYLLACEAE

Arenaria serpyllifolia L. Meadows and dry woods, 5,000 ft. Big Flat (*1141, 1195*).

Silene menziesii Hook. 5,000 ft. South Fork of Salmon R. near Big Flat (*H13,184A, H13,325*).

var. *viscosa* (Greene) Hitchc. & Maguire. 5,000 ft. South Fork of Salmon R. near Big Flat (*H13,184*).

CHENOPODIACEAE

Chenopodium incognitum H. A. Wahl. 5,000 ft. South Fork of Salmon R. near Big Flat (*H13,316*).

COMPOSITAE

Agoseris heterophylla (Nutt.) Greene. Open meadow, 5,000 ft. Big Flat (*1136*).

A. retrorsa (Benth.) Greene. Loose soil on open slope, 7,000 ft. Packers Pk. (*1154*).

Cirsium vulgare (Savi) Ten. Disturbed areas where logging has occurred, 5,000 ft. Not collected but observed at Big Flat.

Crepis acuminata Nutt. 5,000 ft. South Fork of Salmon R. near Big Flat (*H13,331*).

C. occidentalis Nutt. 6,500–7,500 ft. Caribou Basin (*H13,386*).

Nothocalais alpestris (Gray) Chamb. Loose soil on open slope, 7,000 ft. Packers Pk. (*1156*).

CRASSULACEAE

Sedum purdyi Jeps. Canyon wall, 5,000 ft. Between Big Flat and Caribou Gulch (*H13,565*).

CRUCIFERAE

Arabis glabra (L.) Bernh. 5,000 ft. Big Flat (*H13,201*).

A. repanda Wats. 5,000 ft. Big Flat (*H13,374*).

CUSCUTACEAE

Cuscuta californica H. & A. On *Monardella* and *Sitanion*, rocky, serpentine slope, 7,500 ft. Red Rock Mt. (*1465*).

CYPERACEAE

Carex brainerdii Mkze. 6,500–7,300 ft. Upper end of Stuart Fork (*H13,504*).

C. brevipes W. Boott. 5,000–7,500 ft. South Fork of Salmon R. near Big Flat (*H13,198*), Caribou Basin (*H13,486*).

C. hoodii Boott. Dry sand at meadow edge, 5,000 ft. Big Flat campground (*R. Bacigalupi 7233*).

C. lanuginosa Michx. 5,000 ft. South Fork of Salmon R. near Big Flat (*H13,333*).

C. leptopoda Mkze. 5,000 ft. South Fork of Salmon R. near Big Flat (*H13,197*).

C. microptera Mkze. 5,000 ft. South Fork of Salmon R. near Big Flat (*H13,252*).

C. multicaulis Bailey. Open woods, 5,000 ft. Big Flat (*1127*), Caribou Gulch (*H13,556*).

EQUISETACEAE

Equisetum laevigatum A. Br. Bogs and wet places, 5,000 ft. Big Flat (*1137*).

ERICACEAE

Gaultheria humifusa (Grah.) Rydb. Moist meadows and banks, 7,000–7,200 ft. Caribou Basin (*Lester Rowntree sans no.*), east of Papoose L. (*J. O. Sawyer 2347*).

FAGACEAE

Quercus chrysolepis Liebm. 5,500 ft. Caribou Gulch (*H13,543*).

Q. kelloggii Newb. Montane Chaparral, 5,000 ft. Stuart Fork (*J. O. Sawyer 2358*).

GRAMINEAE

Bromus vulgaris (Hook.) Shear. 5,000 ft. South Fork of Salmon R. near Big Flat (*H13,364*).

Danthonia californica Bol. var. *americana* (Scribn.) Hitchc. 5,000 ft. Big Flat (*H13,228*).

Melica aristata Thurb. ex Bol. 5,000 ft. South Fork of Salmon R. near Big Flat (*H13,365*).

M. subulata (Griseb.) Scribn. 5,000 ft. South Fork of Salmon R. near Big Flat (*H13,319*).

Muhlenbergia minutissima (Steud.) Swall. Bed of drying lake, 5,000 ft. Big Flat (*H13,218*).

Poa bolanderi Vasey. 5,000 ft. South Fork of Salmon R. near Big Flat (*H13,321*).

P. pratensis L. 5,000 ft. South Fork of Salmon R. near Big Flat (*H13,347*).

Stipa californica Merr. & Davy. 5,000 ft. South Fork of Salmon R. near Big Flat (*H13,258*).

HYDROPHYLLACEAE

Hydrophyllum fendleri (Gray) Heller var. *albifrons* (Heller) Macbr. 5,000 ft. South Fork of Salmon R. near Big Flat (*H13,575*).

Phacelia leonis J. T. Howell. Sandy soil, 6,000 ft. Union L. (*H. M. Hall 8613*).

JUNCACEAE

Juncus hemiendytus F. J. Herm. 5,000 ft. Big Flat (*H13,213*). This is the type collection.

J. sphaerocarpus Nees. 5,000 ft. Big Flat (*H13,231*).

J. tenuis Willd. var. *congestus* Engelm. 5,000 ft. South Fork of Salmon R. near Big Flat (*H13,355*).

LEGUMINOSAE

Trifolium eriocephalum Nutt. 5,000 ft. South Fork of Salmon R. near Big Flat (*H13,360*).

LOASACEAE

Mentzelia dispersa Wats. 5,000 ft. South Fork of Salmon R. near Big Flat (*H13,276*).

MARSILEACEAE

Marsilea vestita Hook. & Grev. 5,000 ft. Big Flat (*H13,215*).

ONAGRACEAE

Epilobium alpinum L. 6,500–7,500 ft. Caribou Basin (*H13,399*).

Gayophytum humile Juss. Meadow, 5,000 ft. Big Flat (*1140, H13,280*).

POLEMONIACEAE

Collomia linearis Nutt. 5,000 ft. South Fork of Salmon R. near Big Flat (*H13,262*). This species has been previously reported for Calif. only from the San Bernardino Mountains, Sierra Nevada, and north to Modoc County (Munz, 1959).

Navarretia divaricata (Torr.) Greene. 5,000 ft. South Fork of Salmon R. near Big Flat (*H13,572*).

POLYGONACEAE

Eriogonum lobbii T. & G. Open forests and rocky places, 6,800–7,500 ft. Caribou Basin (*J. O. Sawyer 2365, H13,420*), Stuart Fork (*H13,506*).

E. marifolium T. & G. Dry, open, rocky to sandy places, 5,600–8,000 ft. Canyon Cr. Lakes (*165*), Boulder Cr. Basin (*1412, 1413*), Caribou Basin (*1378*), Stuart Fork above Mirror L. (*J. O. Sawyer 2368*).

E. spergulinum Gray var. *reddingianum* (Jones) J. T. Howell. 5,000 ft. Salmon R. canyon between Big Flat and Caribou Gulch (*H13,561A*).

Polygonum kelloggii Greene. 5,000 ft. Big Flat (*H13,211*).

PTERIDACEAE

Pellaea brachyptera (T. Moore) Baker. Rocky places, 5,000–7,000 ft. Packers Pk. (*1157*).

RANUNCULACEAE

Delphinium decorum F. & M. ssp. *tracyi* Ewan. South facing talus slope, 6,200 ft. Packers Pk. (*1188*).

Paeonia brownii Dougl. ex Hook. Open, wooded areas, 5,000 ft. Big Flat (*H13,238*).

Ranunculus flammula L. var. *ovalis* (Bigel.) L. Benson. 5,000 ft. Big Flat (*H13,210*).

RHAMNACEAE

Rhamnus californica Esch. ssp. *occidentalis* (Howell) C. B. Wolf. 6,500 ft. Near Dorleska Mine (*H. M. Hall 8604*).

ROSACEAE

Fragaria virginiana L. ssp. *platypetala* (Rydb.) Staudt. 5,000 ft. South Fork of Salmon R. near Big Flat (*H13,361*).

Holodiscus boursieri (Carr.) Rehd. in Bailey. 7,000 ft. Upper end of Stuart Fork (*H13,493*).

Physocarpus capitatus (Pursh) Kuntze. 5,000 ft. South Fork of Salmon R. near Big Flat (*H13,287*).

Rosa pisocarpa Gray. 5,000 ft. South Fork of Salmon R. near Big Flat (*H13,354*).

Sanguisorba annua (Nutt. ex Hook.) T. & G. 5,000 ft. Big Flat (*H13,235*).

RUBIACEAE

Galium aparine L. Rocky opening in woods, 5,000 ft. Big Flat (*1185*).

SALICACEAE

Salix sitchensis Sanson. 6,500–7,500 ft. Caribou Basin (*H13,465*).

SAXIFRAGACEAE

Lithophragma campanulatum Howell. 5,000 ft. South Fork of Salmon R. near Big Flat (*H13,300*).

Philadelphus lewisii Pursh ssp. *californicus* (Benth.) Munz. 5,000 ft. Salmon R. canyon between Big Flat and Caribou Gulch (*H13,560*).

Ribes roezlii Regel. var. *cruentum* (Greene) Rehd. 5,500 ft. Caribou Gulch (*H13,546*).

R. viscosissimum Pursh var. *hallii* Jancz. 5,500–7,500 ft. Caribou Basin (*H13,462*), Caribou Gulch (*H13,549*).

SCROPHULARIACEAE

Mimulus alsinoides Dougl. ex Benth. Shaded, rocky seep, 6,800 ft. Packers Pk. (*1151*).

Verbascum thapsus L. Disturbed places along roadsides and in logged areas, 5,000 ft. Not collected but observed along upper Coffee Cr. road and at Big Flat.

VIOLACEAE

Viola lobata Benth. var. *integrifolia* Wats. South Fork of Salmon R. near Big Flat (*H13,268*).

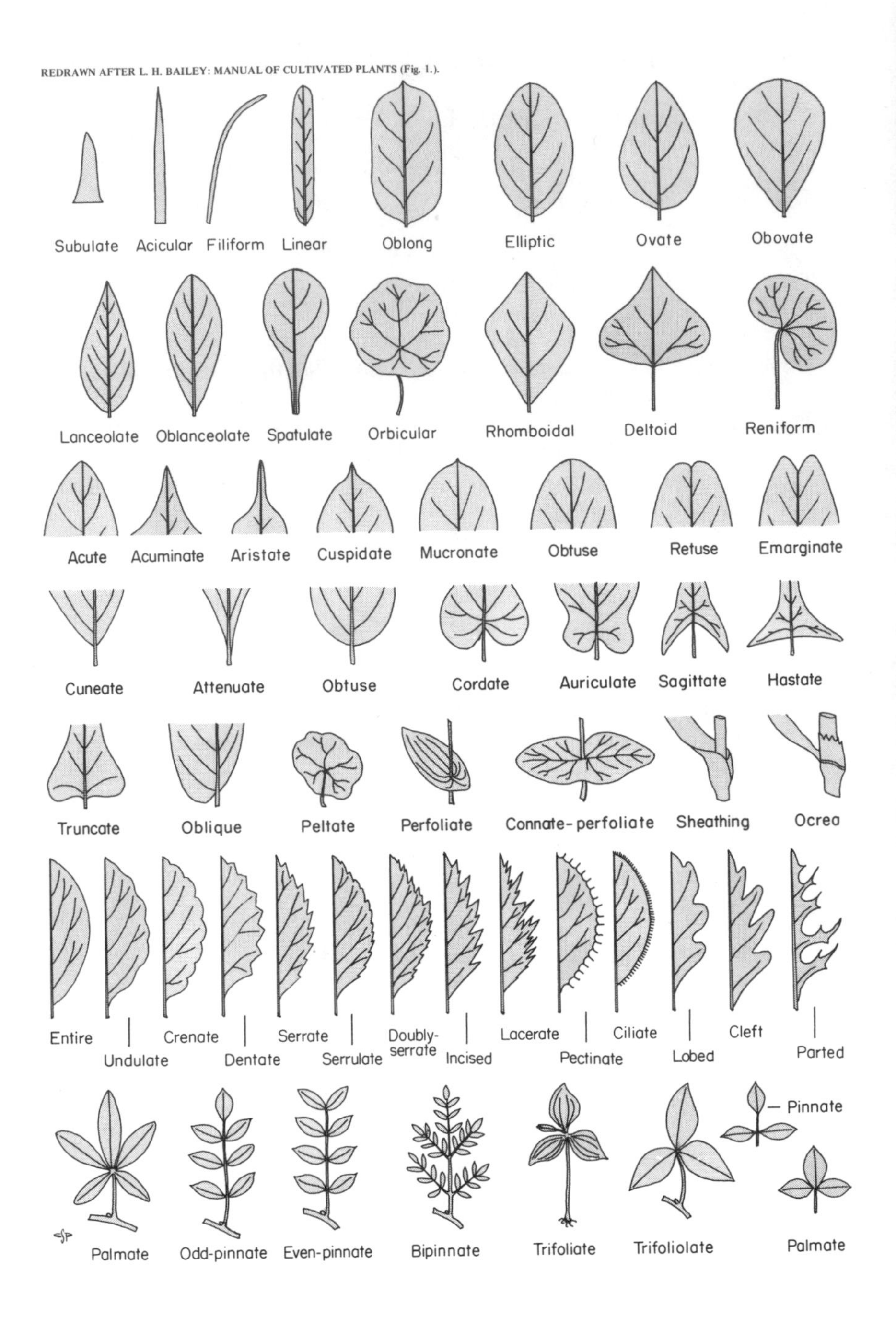

REDRAWN AFTER L. H. BAILEY: MANUAL OF CULTIVATED PLANTS (Fig. 1.).

Glossary

Based on Munz (1959)

ACAULESCENT. Lacking an obvious leafy stem.

ACHENE (akene). A small, dry, hard, indehiscent, 1-seeded fruit; as in the Compositae.

ADNATE. Grown together with an unlike part, as the calyx-tube with an inferior ovary, or an anther by its whole length with the filament.

ANDROGYNOUS. Having staminate flowers above and pistillate flowers below in the same spike or spikelet; as in *Carex*.

ANNUAL. A plant that completes its development from seedling to death in one year or one season.

AURICLE. An ear-shaped appendage. Auriculate: bearing auricles, such as at the base of lvs.

AXIL. Upper angle formed by a leaf or branch with the stem.

AXILLARY. In an axil.

BARBED. Bearing sharp rigid reflexed points like the barb of a fish-hook; as in the pappus bristles of many Compositae.

BARBELLATE. With short, usually stiff hairs.

BRACT. A reduced leaf subtending a flower, usually associated with an inflorescence. Bracteate: with bracts.

BRACTLET. A secondary bract borne on a pedicel instead of subtending it; sepaloid organs subtending the sepals in many Rosaceae.

BULBLET. A small bulb, especially one borne aerially as in a leaf axil or in the inflorescence; as in some *Saxifraga*.

CALLUS. The thickened extension at the base of the lemma in some grasses.

CALYX. The external, usually green, whorl of a flower; contrasted with the inner showy corolla.

CAMPANULATE. Bell-shaped.

CANESCENT. Covered with grayish-white or hoary fine hairs.

CAPILLARY. Hairlike; exceedingly slender.

CARUNCLE. An excrescense or outgrowth at or near the hilum of certain seeds.

CATKIN. A scaly deciduous spike, with the flower parts often much reduced; as in *Salix*.

CAUDATE. Bearing a tail or slender taillike appendage.

CAUDEX. The woody base of an otherwise herbaceous perennial.

CAULINE. Belonging to the stem; as cauline leaves.

CHARTACEOUS. With the texture of writing paper.

CLASPING. A leaf base partly or wholly surrounding the stem; amplexicaul.

CLAVATE. Club-shaped; gradually thickened toward the apex from a slender base.

COMA. A tuft of hairs, particularly on a seed; as in *Asclepias*.

COMOSE. Furnished with a coma.

CONFLUENT. Blending of one part into another.

CONNATE. Congenitally united, as petals joined to form a tube. Connate-perfoliate: united at base in pairs around the supporting axis.

CONNIVENT. Converging or coming together, but not organically united.

CORIACEOUS. Leathery in texture; tough.

COROLLA. The inner whorl of floral leaves, composed of colored petals, which may be almost wholly united or completely separate.

CORYMB. A flat-topped or convex racemose flower-cluster, the lower or outer pedicels longer, their flowers opening first.

CROWN. The persistent base of an herbaceous perennial.

CULM. The type of hollow or pithy, slender stem found in grasses and sedges.

CYATHIUM. A cuplike involucre enveloping the reduced unisexual flowers of *Euphorbia*.

CYME. A flat-topped or convex paniculate flower-cluster, with central flowers opening first. Cymose: arranged in cymes.

DECUMBENT. With the base prostrate but the upper parts erect or ascending.

DEHISCENT. Opening spontaneously when ripe to discharge the contents; as an anther or seed capsule.

DENTICULATE. Slightly and finely toothed.

DIFFUSE. Scattered; widely spread.

DILATED. Flattened and broadened, as an expanded filament or bristle.

DIOECIOUS. Having male and female flowers on different plants.

DISCOID. In the Compositae, a head without ray-florets.

DISK. Flowers or corollas. The central, tubular corollas of a head in the Compositae.

DISSECTED. Deeply divided into numerous fine segments.

DIVARICATE. Widely divergent or spreading.

DRUPE. A fleshy one-seeded indehiscent fruit containing a stone with a kernel; a stone-fruit such as a plum.

EQUITANT. Astride, as if riding, such as the leaves of an Iris.

EROSULATE, EROSE. Irregularly toothed as if gnawed.

EXSERTED. Protruding, as stamens projecting beyond the corolla; not included.

EXTRA-AXILLARY. Borne on the stem above or below the leaf axils.

FALCATE. Sickle-shaped.

FASCICLE. A close cluster or bundle of flowers, leaves, stems, or roots.

FILIFORM. Threadlike.

FLABELLATE, FLABELLIFORM. Fan-shaped; broadly wedge-shaped.

FLOCCOSE. With tufts of soft woolly hair. Flocculent: the diminutive.

FLORET. The individual flower of the Compositae and Gramineae; a small flower of a dense cluster.

FOLIOLATE. Having leaflets.

FOLLICLE. A dry fruit of one carpel, opening only on the ventral suture; as in *Delphinium*.

FUSIFORM. Spindle-shaped; thickest near the middle and tapering toward each end.

GALEA. The helmetlike upper lip in certain bilabiate corollas; as in Scrophulariaceae.

GALEATE. Having a galea.

GENICULATE. Bent abruptly, as a knee; such as the awns in some grasses.

GLABRATE. Almost glabrous; tending to be glabrous. Glabrescent: becoming glabrous.

GLABROUS. Without hairs.

GLAUCOUS. Covered or whitened with a bloom, as a cabbage leaf.

GLOBOSE. Spherical or rounded. Globular: somewhat or nearly globose.

GLUMES. The pair of bracts at the base of a grass spikelet.

GYNAECANDROUS. Having staminate flowers below and pistillate flowers above in the same spike or spikelet.

HERB. A plant without persistent woody stem, at least above ground.

HIRSUTE. Rough with coarse or shaggy hairs.

HISPID. Rough with stiff or bristly hairs.

HOARY. Covered with white down. Appears pale because of dense, whitish, pubescence.

HYALINE. Colorless or translucent, transparent.

IMBRICATE. Overlapping as shingles on a roof; as the phyllaries in *Aster*.

INDUSIUM. In ferns, the epidermal outgrowth that covers or invests the sorus.

INFLORESCENCE. The flower-cluster of a plant; arrangement of flowers on stem.

INTERNODES. The portion of stem between two nodes.

INVOLUCRE. A whorl of bracts (phyllaries) subtending a flower cluster; as in the heads of Compositae.

IRREGULAR. Asymmetric, as a zygomorphic flower; with only one plane of symmetry.

LACINIATE. Cut into narrow lobes or segments.

LATERAL. At or on the side.

LAX. Loose, distant.

LYRATE. Lyre-shaped.

MONOECIOUS. Having male and female flowers on the same plant but lacking perfect ones.

MUCRO. A small and short, abrupt tip of an organ, as the projection of the midrib of a leaf. Mucronate: with a mucro.

NERVE. A simple vein or slender rib of a seed or bract; as on the glumes and lemmas of grasses.

OBSOLETE. Rudimentary or not evident; vestigial.

OVOID. Solid ovate or solid oval; as a hen's egg.

PALEA. One of the chafflike scales on the receptacle of many Compositae; the inner bract of a grass floret, often partly invested by the lemma.

PALEACEOUS. Chaffy; composed of small membranaceous scales.

PALMATE. Hand-shaped with the fingers spread; in a leaf, having the lobes or divisions radiating from a common point.

PANICLE. A compound racemose inflorescence; usually with many branches.

PANICULATE. Borne in a panicle.

PAPILLATE, PAPILLOSE. Bearing minute conical processes or papillae.

PAPPUS. The modified calyx in Compositae, consisting of a crown of bristles or scales on the summit of the achene.

PEDICEL. The stalk of a single flower in a flower-cluster or a spikelet in grasses.

PEDUNCLE. The general term for the stalk of a flower or a cluster of flowers. Pedunculate: Having a peduncle.

PENDENT, PENDULOUS. Suspended or hanging; nodding.

PERENNIAL. Lasting from year to year.

PERFECT (flowers). A flower having both male and female parts (stamens and pistils).

PERIANTH. The floral envelopes collectively; usually used when calyx and corolla are not clearly differentiated; as in *Polygonum* or *Allium*.

PERIGYNIUM. The inflated saclike organ surrounding the pistil in *Carex*.

PETAL. One of the leaves or parts of a corolla, usually colored.

PETALOID. Having the aspect of a colored petal.

PETIOLATE. Having a petiole.

PETIOLE. The stalk of a leaf.

PHYLLARY. An individual bract of the involucre in the Compositae.

PILOSE. Bearing soft and straight spreading hairs.

PINNATE. A compound leaf, having the leaflets arranged on each side of a common petiole; featherlike. Odd-pinnate: pinnate with a single terminal leaflet. Even-pinnate: pinnate without an odd terminal leaflet.

PINNATIFID. Pinnately cleft into narrow lobes not reaching to the midrib.

PISTIL. The ovule-bearing organ of a flower, consisting of stigma and ovary, usually with a style between; gynoecium.

PLUMOSE. Feathery; having fine hairs on each side of a bristle, as a plume.

PROCUMBENT. Trailing on the ground, but not rooting.

PROSTRATE. Lying flat upon the ground.

PUBERULENT. Minutely pubescent.

PUBESCENT. Covered with short soft hairs; downy.

PUNCTATE. Dotted with punctures or with translucent pitted glands.

RACEME. A simple, elongated, indeterminate inflorescence with each flower subequally pedicelled.

RACEMOSE. Of the nature of a raceme or in racemes.

RACHILLA. A small rachis, specifically the axis of a grass spikelet.

RACHIS. The axis of a spike or raceme, or of a compound leaf.

RADIATE. In the Compositae, a head with marginal strap-shaped flowers; having ray flowers.

RAY. A primary branch of an umbel; the ligule of a ray-floret in Compositae, the ray-florets being marginal and differentiated from the disk-florets.

RECEPTACLE. That portion of the floral axis upon which the flower parts are borne, or, in Compositae, that which bears the florets in the head.

RECURVED. Bent backwards.

REFLEXED. Abruptly bent downward.

REGULAR. A flower having radial symmetry, with the parts in each series alike; having 2 or more planes of symmetry.

REMOTE. Distantly spaced.

RETRORSE. Bent backward or downward.

RHIZOME. An underground stem or rootstock, with scales at the nodes and producing leafy shoots on the upper side and roots on the lower side. Rhizomatous: having a rhizome.

RIB. The primary vein of a leaf, or a ridge on a fruit.

RIBBED. With prominent ribs.

ROOTSTOCK. See RHIZOME.

ROSETTE. A crowded cluster of radiating basal leaves.

SACCATE. Furnished with a sac or pouch.

SCABROUS. Rough to the touch, owing to the structure of the epidermis or to the presence of short stiff hairs. Scabrid: somewhat rough.

SCARIOUS. Thin, dry, and membranaceous, not green.

SCROTIFORM. Pouch-shaped.

SEEP(S). A wet spot where underground water comes to or near the surface.

SEPAL. A leaf or segment of the calyx. Sepaloid: sepallike.

SERIATE. Disposed in series or rows; as the phyllaries in some Compositae.

SESSILE. Attached directly by the base; not stalked, as a leaf without a petiole.

SETA. A bristle, or a rigid, sharp-pointed, bristlelike organ. Setaceous: bristly or bristlelike.

SHEATH. The tubular basal part of the leaf that encloses the stem; as in grasses and sedges.

SHRUB. A woody plant of smaller proportions than a tree, which usually produces several branches from the base.

SIGMOID. Doubly curved, like the letter S.

SILICLE. A short silique, not much longer than wide.

SILIQUE. A narrow many-seeded capsule of the Cruciferae, several times longer than wide.

SINUS. The cleft or recess between two lobes of an expanded organ such as a leaf or anther.

SORDID. Of a dull or dirty hue.

SPIKE. An elongated rachis of sessile flowers or spikelets.

SPREADING. Diverging almost to the horizontal; nearly prostrate. Spreading hairs; not at all appressed, but erect. Spreading lower lip; diverging from the main axis of the flower.

SPUR. A slender, saclike, often nectar producing process from a petal or sepal.

STAMEN. The male organ of the flower, consisting of filament and anther which bears the pollen.

STAMINODE. A sterile stamen (lacking an anther); as in *Penstemon.*

STELLATE. Star-shaped.

STERILE. Infertile or barren, as a stamen lacking an anther, a flower wanting a pistil, a seed without an embryo, etc.

STIGMA. The receptive part of the pistil on which the pollen germinates. Stigmatic: pertaining to the stigma.

STIPULE. One of the pair of usually bract-like appendages found at the base of the petiole in many plants. Stipulate: possessing stipules.

STOLON. A modified stem bending over and rooting at the tip; or creeping and rooting at the nodes; or a horizontal stem that gives rise to a new plant at its tip. Cf. Runner, a very slender stolon, and Rhizome, a subterranean stem. Stoloniferous: having stolons.

STRIATE. Marked with fine longitudinal lines or furrows.

STRIGOSE. Clothed with sharp and stiff appressed straight hairs.

STYLE. The stalklike part of some pistils between the ovary and the stigma. Style branches may be only in part stigmatic, the remainder then being an appendage.

SUBTEND. To be below and close to, as the leaf subtends the shoot borne in its axil.

SUFFRUTESCENT. Obscurely shrubby; very little woody, but not necessarily low.

SUTURE. The line of dehiscence of fruits or anthers.

TAPROOT. A primary stout vertical root giving off small laterals but not dividing.

TAWNY. Dull brownish-yellow; fulvous.

TERNATE. In threes, as a leaf consisting of three leaflets. Biternate: twice divided in three.

TOMENTOSE. With tomentum; covered with a rather short, densely matted, soft white wool.

TOMENTUM. A covering of densely matted woolly hairs.

TOOTH. Any small marginal lobe. Toothed: dentate.

TOROSE. Cylindrical with alternate swellings and constrictions; as in some seed capsules.

TORULOSE. The diminutive.

TRIFID. Three-cleft to about the middle.

TUBE. The narrow basal portion of a corolla or calyx whose parts are united.

TURBINATE. Top shaped; inversely conical.

UMBEL. A flat or convex flower-cluster in which the pedicels arise from a common point, like rays of an umbrella. Umbellate: borne in an umbel. Umbellet: a secondary umbel. Umbelliferous: bearing umbels. Umbelliform: umbel-shaped.

UNCINATE. Hooked at the tip.

URCEOLATE. Urn-shaped or pitcherlike, contracted at the mouth.

VILLOUS. Bearing long and soft and not matted hairs; shaggy.

VISCID, VISCOUS. Sticky; glutinous.

WING. A thin and usually dry extension bordering an organ such as a petiole or seed; a lateral petal of a pea flower. Winged: bearing a wing.

WOOLLY. Having long, soft, entangled hairs; lanate. Cf. Tomentose.

ZYGOMORPHIC, ZYGOMORPHOUS. Bilaterally symmetrical; that which can be bisected by only one plane into similar halves. Cf. Irregular.

Bibliography

ABRAMS, LEROY. 1940. *Illustrated flora of the Pacific states.* Vol. I, Ophioglossaceae to Aristolochiaceae. Stanford University Press, Stanford. x + 538 pp.

———. 1944. *Illustrated flora of the Pacific states.* Vol. II, Polygonaceae to Krameriaceae. Stanford University Press, Stanford. viii + 635 pp.

———. 1951. *Illustrated flora of the Pacific states.* Vol. III, Geraniaceae to Scrophulariaceae. Stanford University Press, Stanford. viii + 866 pp.

ABRAMS, LEROY and FERRIS, ROXANA. 1960. *Illustrated flora of the Pacific states.* Vol. IV, Bignoniaceae to Compositae. Stanford University Press, Stanford. v + 732 pp.

CHAMBERS, KENTON L. 1963. *Claytonia nevadensis* in Oregon. Leafl. W. Bot. 10 (1):1–6.

DAVIS, G. A., HOLDAWAY, M. J., LIPMAN, P. W. and ROMEY, W. D. 1965. Structure, metamorphism and plutonism in the south-central Klamath Mountains, California. Geol. Soc. Am. Bull. 76(8):933–965.

DEMPSTER, LAURAMAY, and EHRENDORFER, F. 1965. Evolution of the *Galium multiflorum* complex in western North America. II. Critical taxonomic revision. Brittonia 17(4):289–334.

EASTWOOD, ALICE. 1902. From Redding to the snow clad peaks of Trinity County; also, list of trees and shrubs en route. Sierra Club Bull. 4:39–58.

FERLATTE, WILLIAM J. 1972. *Haplopappus lyallii* (Compositae), a new record from California. Madroño 21 (8):535.

GENTRY, J. L. 1972. A new combination and a new name in *Hackelia* (Boraginaceae). Madroño 21 (7):490.

HINDS, N. E. 1952. Evolution of the California landscape. Calif. Div. of Mines Bull. 158:139–142.

HITCHCOCK, A. S. 1950. *Manual of the grasses of the United States.* U. S. Dept. of Agriculture, Misc. Pub. No. 200. Washington, D.C. 1051 pp.

HITCHCOCK, C. L., CRONQUIST, A., OWENBY, M. and THOMPSON, J. W. 1969. *Vascular plants of the Pacific Northwest.* University of Washington Press, Seattle and London. 1:61–63.

IRWIN, W. P. 1960. Geologic reconnaissance of the northern Coast Ranges and Klamath Mountains, California. Calif. Div. of Mines Bull. 179. 80 pp.

KECK, DAVID. 1940. Studies in *Penstemon* VII. Amer. Midl. Naturalist 23: 603–605.

LIPMAN, P. W. 1962. Geology of the southwestern Trinity Alps, northern California. Stanford University Ph.D. Thesis. University Microfilms, Inc. Ann Arbor, Mich. xvi + 210 pp.

MacGinitie, Harry D. 1937. The flora of the Weaverville Beds of Trinity County, California, with descriptions of the plant bearing beds. Reprinted from Carnegie Institution of Washington Publ. No. 465, pp. 83–151.

Munz, Philip A. 1959. *A California flora.* University of California Press, Berkeley and Los Angeles. 1681 pp.

———. 1968. *Supplement to a California flora.* University of California Press, Berkeley and Los Angeles. 224 pp.

Munz, P. A. and Thorne, R. F. 1973. A new California *Trillium.* Aliso 8(1): 15–17.

Sharp, R. P. 1960. Pleistocene glaciation in the Trinity Alps of northern California. Amer. J. Sci. 258:305–340.

Wilson, F. Douglas. 1963. Revision of *Sitanion* (Triticeae, Gramineae). Brittonia 15:303–323.

Index to English Names

Index to Latin Names